Ergebnisse der Mathematik und ihrer Grenzgebiete

Band 35

Herausgegeben von
P. R. Halmos · P. J. Hilton · R. Remmert · B. Szőkefalvi-Nagy
Unter Mitwirkung von
L. V. Ahlfors · R. Baer · F. L. Bauer · R. Courant · A. Dold · J. L. Doob
S. Eilenberg · M. Kneser · H. Rademacher · B. Segre · E. Sperner
Redaktion: P. J. Hilton

Calculus of Fractions and Homotopy Theory

P. Gabriel and M. Zisman

Springer-Verlag Berlin Heidelberg GmbH 1967

Professor Dr. Peter Gabriel
Professor Dr. Michel Zisman
Université de Strasbourg
Departement de Mathématique Strasbourg

ISBN 978-3-642-85846-8 ISBN 978-3-642-85844-4 (eBook)
DOI 10.1007/978-3-642-85844-4

Originally published by Springer-Verlag, Berlin · Heidelberg 1967
Softcover reprint of the hardcover 1st edition 1967

Library of Congress Catalog Card Number 67-10470

Title-No. 4579

Introduction

The main purpose of the present work is to present to the reader a particularly nice category for the study of homotopy, namely *the homotopic category* (IV). This category is, in fact, — according to Chapter VII and a well-known theorem of J. H. C. WHITEHEAD — equivalent to the category of CW-complexes modulo homotopy, i.e. the category whose objects are spaces of the homotopy type of a CW-complex and whose morphisms are homotopy classes of continuous mappings between such spaces. It is also equivalent (I, 1.3) to a category of fractions of the category of topological spaces modulo homotopy, and to the category of Kan complexes modulo homotopy (IV).

In order to define our homotopic category, it appears useful to follow as closely as possible methods which have proved efficacious in homological algebra. Our category is thus the "topological" analogue of the derived category of an abelian category (VERDIER).

The algebraic machinery upon which this work is essentially based includes the usual grounding in category theory — summarized in the Dictionary — and the theory of *categories of fractions* which forms the subject of the first chapter of the book. The merely topological machinery reduces to a few properties of *Kelley spaces* (Chapters I and III).

The starting point of our study is the category $\Delta^\circ\mathscr{E}$ of simplicial sets (C.S.S. complexes or semi-simplicial sets in a former terminology). Notwithstanding a very large number of papers and seminar notes published on the subject, no book has yet been devoted to them. Therefore in order to fill this gap to some extent, we go back to the beginning of the theory, and give a complete proof of theorems well-known to the specialist, in the hope of providing the reader with a coherent survey, and presenting some proofs which are easier or more conceptual than those already published.

This book is thus intended to appeal at the same time to the beginner who wishes to learn algebraic topology, to the algebraist who wants to be acquainted with topology, and to the topologist eager to assimilate the category language.

Such a program, which *a priori* seems very ambitious, is in fact very limited: it cannot be greater than the number of pages in the volumes where this work has been published. Thus the point where we leave off is, in fact, nothing but the starting point of algebraic topology, and this book is thus only an introduction to that theory.

Let us summarize briefly the content of our work.

Chapter I sets forth the theory of categories of fractions and gives a few examples of applications to groupoids, Kelley spaces and abelian categories. Given a category $\mathscr{C}$ and a subset Σ of the set $\mathscr{A}r\,\mathscr{C}$ of the morphisms of $\mathscr{C}$, a category $\mathscr{C}[\Sigma^{-1}]$ is constructed whose objects are the same as those of $\mathscr{C}$, but where the morphisms of Σ have been formally made invertible. The description of the set $\mathscr{A}r\,\mathscr{C}[\Sigma^{-1}]$ is particularly nice when Σ possesses some properties "*allowing a calculus of fractions*" since in that case any morphism of $\mathscr{C}[\Sigma^{-1}]$ can be written $s^{-1}f$ where s is in Σ and f in $\mathscr{A}r\,\mathscr{C}$. The interest of this concept lies mainly in its relationship to the existence of adjoint functors (I, 1.3 and I, 4.1).

After having recalled a few properties of the category of functors with values in a set, Chapter II gives the definition of the category $\Delta^{\circ}\mathscr{E}$ of simplicial sets, and draws the first inferences from it. One constructs a fully faithful functor from the category $\mathscr{C}at$ into $\Delta^{\circ}\mathscr{E}$ which has a left adjoint. This pair of adjoint functors allows us to define certain other pairs of adjoint functors, and in particular the pair (Π, D) where ΠX is the *Poincaré groupoid* of the simplicial set X, and DG is a $K(\Pi_1 G, 1)$-complex where $\Pi_1 G$ is the Poincaré group of the groupoid G. This concept allows us finally to construct an extremely simple theory for the fundamental group of a pointed simplicial set, and in particular to state a *Van Kampen theorem* in the category $.\Delta^{\circ}\mathscr{E}$.

Chapter III is concerned with the study of the functor $|?|$, that is, MILNOR's geometric realization functor. After having shown that the geometric realization of a simplicial set has some good properties (it is a Hausdorff space, locally arcwise connected and locally contractible), it is proved that the functor $|?|$ has some interesting exactness properties too, and that it commutes with locally trivial morphisms *provided one considers the range of $|?|$ to be the category of Kelley spaces* instead of the whole category of topological spaces. Under this new definition, the geometric realization functor commutes with direct limits and finite inverses limits. Moreover it transforms a locally trivial morphism into a Serre fibration.

With Chapter IV the study of homotopy begins. After having defined the homotopy relation between morphisms without any restriction (and not only when the common range is a Kan complex), the category $\overline{\Delta^{\circ}\mathscr{E}}$ of complexes modulo homotopy and a special set of arrows in that category — *the anodyne extensions* —, it only remains for us to define the homotopic category $\mathscr{H}$ as the category of fractions of $\overline{\Delta^{\circ}\mathscr{E}}$ where the anodyne extensions are made invertible. Since, for any simplicial set X, there exists an anodyne extension a_X: $X \to X_K$ where X_K is a Kan complex, the category $\mathscr{H}$ is equivalent to the category of Kan complexes modulo homotopy. Chapter IV also gives a variant "with base points" of the preceding theory, and contains a few technical results on Kan fibrations and Kan complexes. It is finally pointed out — as an exercise — how the Π-theory given in Chapter II fits into this new context.

Chapter V is independent of the preceding ones and presupposes only a few elementary results (recalled in the Dictionary and in Chapter I) about groupoids. Its purpose is to give a standard and self-dual proof of various exact sequences occuring in algebraic topology. As an example of possible applications, the reader will find the proof of a few well known exact sequences (PUPPE, ECKMANN-HILTON); he will be able to obtain in the same way all the other exact sequences of ECKMANN-HILTON [*1*]. The main idea is to construct an exact sequence in the 2-category of pointed groupoids and then to reduce the study of a large class of 2-categories to the preceding one.

We should point out that the preceding method allows to give an easy proof of the exactness of the sequence of SPANIER-WHITEHEAD in S-theory[1].

Chapter VI is chiefly an application of the preceding chapter to simplicial sets and to the homotopic category. It also gives the definition of homotopy groups, and various technical developments concerning *minimal fibrations*, whose purpose is to prove (i) that every fibration is homotopically equivalent (modulo the base) to a locally trivial morphism, and (ii) the J. H. C. Whitehead theorem for simplicial sets.

Finally Chapter VII is limited to bringing together the preceding material in order to prove the theorems referred to in the beginning of this introduction, and which constitute the justification of the work itself.

All this, as has already been said, is only an introduction to algebraic topology. To make this introduction at least more or less complete, we have sketched briefly in two appendices a few complementary remarks of interest to the reader.

In Appendix I will be found a theory of coverings and local systems, and as an application a proof of the Van Kampen theorem for the geometric realization of simplicial sets.

In Appendix II the reader will find, as a bonus, a version of EILENBERG's theorem connecting the homology of a complex with the singular

[1] The theory of carriers and S-theory — Algebraic geometry and Topology. A symposium in honor of S. LEFSCHETZ. Princeton University Press, 1956, pp. 330—360.

homology of its geometrical realization, and the spectral sequence of a fibration.

The present volume is an outgrowth of a seminar given by the authors in 1963/64 in the *Institut de Mathématique de Strasbourg*, with the help of C. GODBILLON.

We wish to thank Professor A. DOLD who asked us to write this book for Springer-Verlag, and Professor P. HILTON who read the first draft of this book and accepted it for publication in the *Ergebnisse* series. We thank also Mr. LUC DEMERS qui a été chargé de la tâche ingrate de traduire le manuscript dans la langue d'outre-Manche.

Strasbourg, 1. 6. 1966 P. GABRIEL, M. ZISMAN

Contents

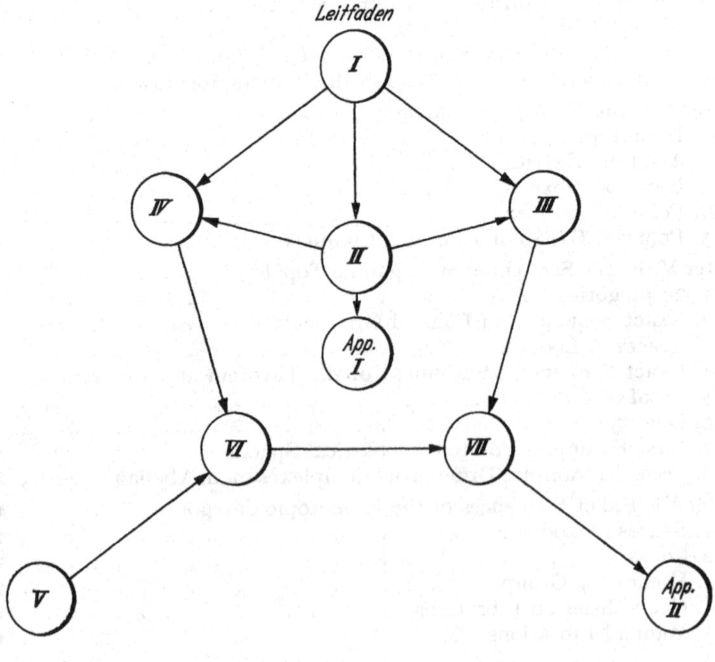
Leitfaden
I
IV
II
III
App.
I
V
VI
VII
App.
II

Dictionary

The aim of this dictionary is to define with precision the terms which will be used in the sequel. For the basic notions, we refer the reader to the following works:

GABRIEL, P.: Catégories abéliennes. Bull. Soc. Math. France **90** (1962). GROTHENDIECK, A.: Sur quelques points d'algèbre homologique. Tohoku math. J. serie 2. **9** (1957). MACLANE, S.: Homology. Berlin-Heidelberg-New York: Springer. MITCHELL, B.: Theory of Categories. New York: Academic Press.

Unfortunately, the terminologies used in these books coincide neither with each other, nor with those which we will sometimes use. It is this great variety of language which forces us to restrict the number of publications given as references.

Adjoint: See GABRIEL (op. cit.) for the notations. We say that T is left adjoint to S and that S is right adjoint to T. We say that ψ is an adjunction isomorphism from T to S, that Ψ is an adjunction morphism from T to S; similarly, φ is an adjunction isomorphism from S to T, and Φ is an adjunction morphism from S to T.

We say that Ψ is quasi inverse to Φ, and conversely.

Amalgamated sum: It is equivalent to the expression "somme fibrée" of GABRIEL (loc. cit.) and pushout of MITCHELL (loc. cit.). We write $A \overset{C}{\amalg} B$ or $A \overset{a,b}{\amalg} B$ for the amalgamated sum of a diagram of the form

$$A \xleftarrow{a} C \xrightarrow{b} B.$$

Arrow: See category and diagram scheme.

Can: Short for canonical.

Category: See the references. If $\mathscr{C}$ is a category, we will write $\mathfrak{Ob}\,\mathscr{C}$ (resp. $\mathfrak{Ar}\,\mathscr{C}$) for the class of its objects (resp. morphisms or arrows). The *identity morphism* of an object c of a category $\mathscr{C}$ will be denoted by $\mathrm{Id}_{\mathscr{C}}\, c$, or simply by $\mathrm{Id}\, c$. If f is a morphism of a category C, the *domain* and the *range* of f will be denoted by $\mathfrak{d}_{\mathscr{C}} f$ and $\mathfrak{r}_{\mathscr{C}} f$, or simply by $\mathfrak{d} f$ and $\mathfrak{r} f$. The set of morphisms of a category $\mathscr{C}$ with domain a and range b will be denoted by $\mathscr{C}(a, b)$, or $\mathrm{Hom}_{\mathscr{C}}(a, b)$.

Category of paths of a diagram scheme T: It is a category $\mathscr{P}a\,T$ whose objects are the same as those of T, and whose morphisms are the sequences $(a_1, a_2, \ldots a_n)$ of morphisms of T such that $\mathfrak{r}\, a_i = \mathfrak{d}\, a_{i+1}$ $(i=1, 2, \ldots n-1)$. If d_T is the diagram of $\mathcal{P}a\, T$ of type T which induces the identity on objects and which associates with each morphism a of T the sequence formed by the unique morphism a, the pair $(\mathcal{P}a\, T, d_T)$ can be characterized by the following property: for each category $\mathcal{C}$ and each diagram e of $\mathcal{C}$ of type T, there is one and only one functor $E\colon \mathcal{P}a\, T \to \mathcal{C}$ such that $E\, d_T = e$.

$\mathcal{C}at$: It is the *category of categories*, i.e. the category whose objects are the categories $\mathcal{C}$ such that $\mathfrak{Ob}\,\mathcal{C}$ and $\mathfrak{Ar}\,\mathcal{C}$ belong to a given fixed universe. Morphisms of $\mathcal{C}at$ are functors. If x and y are categories, we will write $\mathcal{H}om(x, y)$ for the category whose objects are the functors from x to y, and whose morphisms are the functor morphisms.

Cartesian, Cocartesian: Let

$$\begin{array}{ccc} A & \xrightarrow{u} & B \\ {\scriptstyle x}\downarrow & & \downarrow{\scriptstyle y} \\ C & \xrightarrow[v]{} & D \end{array}$$

be a commutative square of a category $\mathcal{C}$ (i.e. $vx = yu$). We say that the square is *cartesian* if, for each pair (b, c) of morphisms $b\colon E \to B$ and $c\colon E \to C$ such that $vc = yb$, there is one and only one morphism $a\colon E \to A$ such that $b = ua$ and $c = xa$. We say that the square is *cocartesian* if, for each pair (b', c') of morphisms $b'\colon B \to E'$ and $c'\colon c \to E'$ such that $b'u = c'x$, there is one and only one morphism $d\colon D \to E'$ such that $b' = dy$ and $c' = dv$.

Cokernel: Dual to kernel; called coequalizer in MITCHELL (op. cit.).

Cofree: Let $\mathcal{A}$ and $\mathcal{B}$ be two categories, $F\colon \mathcal{A} \to \mathcal{B}$ a functor, and b an object of $\mathcal{B}$. We say that $f\colon Fa \to b$ is a *coliberty* morphism (or right liberty) over b and that a is cofree (or right free) on b if, for each morphism $f'\colon Fa' \to b$, there is one and only one morphism $g\colon a' \to a$ such that $f' = f \cdot (Fg)$.

Conservative: Let $\mathcal{A}$ and $\mathcal{B}$ be categories, $F\colon \mathcal{A} \to \mathcal{B}$ a functor, and f a morphism of A. We say that F is *conservative* if f is invertible if and only if Ff is invertible.

Diagram: Let T and U be two diagram schemes. A *diagram* of U of type T is defined by two maps $d_{\mathfrak{Ar}}\colon \mathfrak{Ar}\, T \to \mathfrak{Ar}\, U$ and $d_{\mathfrak{Ob}}\colon \mathfrak{Ob}\, T \to \mathfrak{Ob}\, U$ such that $\mathfrak{d}_U\, d_{\mathfrak{Ar}} = d_{\mathfrak{Ob}}\, \mathfrak{d}_T$ and $\mathfrak{r}_U\, d_{\mathfrak{Ar}} = d_{\mathfrak{Ob}}\, \mathfrak{r}_T$. When $\mathcal{C}$ is a category, a diagram of $\mathcal{C}$ of type T is a diagram of type T of the diagram scheme underlying to $\mathcal{C}$.

Diagram scheme: A diagram scheme T is given by two sets $\mathfrak{Ob}\, T$ and $\mathfrak{Ar}\, T$ and two maps $\mathfrak{d}_T, \mathfrak{r}_T\colon \mathfrak{Ar}\, T \to \mathfrak{Ob}\, T$; an element of $\mathfrak{Ob}\, T$ is an object, an element of $\mathfrak{Ar}\, t$ is a morphism or an arrow. If a is a morphism, $\mathfrak{d}_T\, a$ is the domain of a, $\mathfrak{r}_T$ is the range of a.

With each category $\mathscr{C}$ is associated a subordinate diagram scheme, which has the same objects and the same domain and range maps. Then the category $\mathscr{C}$ is determined by the underlying diagram scheme and the composition laws.

Domain: See Category.

$\mathscr{E}$: Category of sets.

Epimorphism: See GROTHENDIEK (op. cit.).

Equivalences of categories: See GABRIEL (op. cit.): quasi inverse.

Fibred product: See GABRIEL (op. cit.); called pullback in MITCHELL (op. cit.). We shall write $A \underset{C}{\times} B$ or $A \underset{C}{\prod} B$, $A \underset{f,g}{\prod} B$ for the fibred product of $A \xrightarrow{f} C \xleftarrow{g} B$.

Free: (or left free); liberty morphism (or left liberty): dual to cofree, and coliberty morphisms.

Full, Fully faithful: D is a *subcategory of* $\mathscr{C}$ if $\mathfrak{Ob}\, \mathscr{D} \subset \mathfrak{Ob}\, \mathscr{C}$, $\mathfrak{Ar}\, \mathscr{D} \subset \mathfrak{Ar}\, \mathscr{C}$, and if the composition laws of $\mathscr{D}$ are induced by those of $\mathscr{C}$. We say that $\mathscr{D}$ is a *full* subcategory of $\mathscr{C}$ if moreover, the equality $\mathscr{D}(d, d') = \mathscr{C}(d, d')$ is satisfied for each pair (d, d') of objects of $\mathscr{D}$; a functor $F\colon \mathscr{C} \to \mathscr{C}'$ is said to be *fully faithful* if F defines an equivalence between $\mathscr{C}$ and a full subcategory of $\mathscr{C}$!

Functor: If $F\colon \mathscr{C} \to D$ is a functor and if c, c' are objects of $\mathscr{C}$, we write $F(c, c')$ for the map from $\mathscr{C}(c, c')$ to $\mathscr{D}(Fc, Fc')$ which is defined by F.

Groupoid: It is a category where all morphisms are invertible

$\mathscr{Gr}$: It is the category of groupoids, i.e. the full subcategory of $\mathscr{Cat}$ whose objects are the groupoids.

Initial: See limit.

Invertible: Let $f\colon a \to b$ be a morphism of a category $\mathscr{C}$. We say that f is *invertible* if there is a morphism $g\colon b \to a$ such that $g \circ f = \mathrm{Id}\, a$ and $f \circ g = \mathrm{Id}\, b$.

Kernel: Let $a \underset{g}{\overset{f}{\rightrightarrows}} b$ be a diagram of a category $\mathscr{C}$. A kernel is an inverse limit of this diagram; it is called equalizer in MITCHELL (op. cit.).

Limits: Let d be a diagram of type T of a category $\mathscr{C}$. If x is an object of $\mathscr{C}$, a *projective or inverse cone* with domain vertex x and base d

is a family (p_t) of morphisms indexed by the objects of T; the domain of p_t is x and its range is $dt = d_{\mathfrak{Ob}}(t)$; it is required also that (p_t) satisfies the relations $p_{t'} = d_{\mathfrak{Ar}}(a) \circ p_t = (d\,a) \circ p_t$ for each morphism $a\colon t \to t'$. Such a cone is called *terminal* if, for each cone (q_t) with domain vertex y and with the same base d, there is one and only one morphism $g\colon y \to x$ such that $q_t = p_t \circ g$ for all t. We say then that x is an *inverse limit* of d (notation $x = \varprojlim d$) and that p_t is the *projection* of index t (notation $p_t = pr_t^d$ or pr_t).

We say that C admits finite inverse limits if $\varprojlim d$ exists whenever $\mathfrak{Ob}\, T$ and $\mathfrak{Ar}\, T$ are finite sets.

Let $F\colon \mathscr{C} \to \mathscr{D}$ be a functor. If $\varprojlim d$ and $\varprojlim F d$ exist, we define a morphism $c\colon F \varprojlim d \to \varprojlim F d$ by means of the equations $pr_t^{Fd} \circ c = F(pr_t^d)$. If c is invertible, we say that F commutes with the inverse limit of d.

For the dual notions, we will use the following terminology: *inductive or direct cone* with range vertex x and base d, *initial* cone, direct limit (notation $\varinjlim d$), *induction* or *canonical injection* of index t (notation in_t^d or in_t).

Let us review briefly the construction of the direct limit of a diagram $d\colon T \to \mathscr{C}at$. Let $\mathscr{E}$ be the category of sets; let $\mathfrak{Ob}\, d\colon T \to \mathscr{E}$ (resp. $\mathfrak{Ar}\, d$) be the diagram which associates with each object t of T the set of objects (resp. of morphisms) of the category $d(t)$. By passing to the direct limit, the domain and range maps of $d(t)$ induce maps

$$\mathfrak{d}, \mathfrak{r}\colon \varinjlim_t \mathfrak{Ar}\, d(t) \rightrightarrows \varinjlim_t \mathfrak{Ob}\, d(t)$$

and define a diagram scheme X whose set of objects is $\varinjlim_t \mathfrak{Ob}\, d(t)$ and whose set of morphisms is $\varinjlim_t \mathfrak{Ar}\, d(t)$. It is clear that X is the direct limit of the diagram schemes subordinated to the categories $d(t)$.

Let $\mathscr{P}a\, X$ be the category of paths of X. The canonical maps from $\mathfrak{Ob}\, d(t)$ and $\mathfrak{Ar}\, d(t)$ to $\varinjlim \mathfrak{Ob}\, d(t)$ and $\varinjlim \mathfrak{Ar}\, d(t)$ induce diagrams $i_t\colon d(t) \to \mathscr{P}a\, X$. There is obviously no reason for these diagrams to be functors: for instance, if α and β are two composable morphisms of $d(t)$, $i_t(\beta) \circ i_t(\alpha)$ can be different from $i_t(\beta \circ \alpha)$; similarly, if a is an object of $d(t)$, $i_t(\mathrm{Id}\, a)$ is in general different from $\mathrm{Id}\, i_t(a)$. It follows that $\varinjlim d$ is the quotient of the category of paths $\mathscr{P}a\, X$ by the relations.

$$i_t(\beta) \circ i_t(\alpha) = i_t(\beta \circ \alpha) \quad \text{and} \quad i_t(\mathrm{Id}\, a) = \mathrm{Id}\, i_t(a).$$

Hence, the set of morphisms of the direct limit d is the quotient of $\mathfrak{Ar}\, \mathscr{P}a\, X$ by the intersection S of all equivalence relations R satisfying the following conditions:

a) R contains $(i_t(\beta)\circ i_t(\alpha), i_t(\beta\circ\alpha))$ and $(i_t(\mathrm{Id}\,a), \mathrm{Id}\, i_t(a))$ for all t, a, β, α;

b) the relation $f \underset{R}{\sim} g$ implies that f has the same domain and the same range as g;

c) the relation $f \underset{R}{\sim} g$ implies that $\alpha\circ f \underset{R}{\sim} \alpha\circ g$ and $f\circ\beta \underset{R}{\sim} g\circ\beta$ whenever these expressions make sense.

The construction of inverse limits is more simple; if $d\colon T\to\mathscr{C}at$ is a small diagram, the set of objects of $\varprojlim d$ is the inverse limit of the sets $\mathfrak{Ob}\, d(t)$; if (x_t), (y_t) and (z_t) are elements of $\varprojlim_t \mathfrak{Ob}\, d(t)$, we have

$$\mathrm{Hom}_{\varprojlim d}\big((x_t), (y_t)\big) = \varprojlim_t \mathrm{Hom}_{d(t)}(x_t, y_t)$$

and the composition maps

$$\mathrm{Hom}_{d(t)}(x_t, y_t)\times\mathrm{Hom}_{d(t)}(y_t, z_t)\to\mathrm{Hom}_{d(t)}(x_t, z_t)$$

induce, by passing to the limit, the composition maps of $\varprojlim d$.

Morphism: See category.

Monomorphism: See GROTHENDIECK (op. cit.).

Quasi inverse: See adjoint.

Quasi filtering: A category $\mathscr{C}$ is called (right-) *quasi filtering* if (i) and (ii) are satisfied: (i) for each pair (a, a') of morphisms with the same domain, there is a pair (b, b') of morphisms with the same range such that $ba = b'a'$; (ii) for each pair (C, C') of morphisms with the same domain and the same range, there is one morphism d such that $dc = dc'$.

Range: See category.

Representable: Let $\mathscr{E}$ be the category of sets, $\mathscr{D}$ an arbitrary category, and $\mathscr{D}^\circ\mathscr{E}$ the category of contravariant functors from $\mathscr{D}$ to $\mathscr{E}$. For each object a of $\mathscr{D}$, $b \rightsquigarrow \mathscr{D}(b, a)$ is a contravariant functor from $\mathscr{D}$ to $\mathscr{E}$: we denote this functor by $h_a^{\mathscr{D}}$ or by $\mathscr{D}(a)$, and we say that it is representable. Thus we define a covariant functor $h^{\mathscr{D}}\colon a \rightsquigarrow \mathscr{D}(a)$ from $\mathscr{D}$ to $\mathscr{D}^\circ\mathscr{E}$.

If $F\colon \mathscr{D}^\circ\to\mathscr{E}$ is an arbitrary functor and a an object of $\mathscr{D}$, we know that there is a "Functorial bijection" from $\mathscr{D}^\circ\mathscr{E}\big(\mathscr{D}(a), F\big)$ onto $F(a)$: this bijection associates with each functor morphism $f\colon \mathscr{D}(a)\to F$, the image λ of the identity $\mathrm{Id}\,a$ of a under the map $f(a)\colon D(a, a)\to F(a)$; we can check easily that f can be reconstructed from λ as follows: for each morphism $\alpha\colon b\to a$, $f(b)(\alpha)$ is the image of λ under the map $F(\alpha)\colon F(a)\to F(b)$. In particular, if F is of the form $\mathscr{D}(a')$, λ is a morphism from a to a', and f is simply $\mathscr{D}(\cdot, \lambda)\colon \mathscr{D}(\cdot, a)\to\mathscr{D}(\cdot, a')$. It follows that the functor $h^{\mathscr{D}}\colon a \rightsquigarrow h_a^{\mathscr{D}}$ is fully faithful and allows us to identify $\mathscr{D}$ with a full subcategory of $\mathscr{D}^\circ\mathscr{E}$.

If F, a, λ are defined as above, and if the associated functor morphism $f\colon \mathcal{D}(a) \to F$ is invertible, we say that the pair (a, λ) is a *representation* of F.

Retraction, section: A retraction (resp. section) of a morphism $v\colon b \to a$ is a morphism $u\colon a \to b$ such that $u \circ v = \mathrm{Id}\, b$ (resp. $v \circ u = \mathrm{Id}\, a$).

Small diagram: A diagram d of type T of a category $\mathcal{C}$ is called a *small diagram* if $\mathfrak{Ob}\, T$ and $\mathfrak{Ar}\, T$ belong to a given fixed universe $\mathfrak{U}$. When the sets $\mathcal{C}(x, y)$ also belong to $\mathfrak{U}$, and each small diagram has a direct limit (resp. inverse limit), we say that $\mathcal{C}$ is a *category with direct limits* (resp. inverse limits).

Subcategory: See Full.

Terminal: See limits.

Zero: See GABRIEL (op. cit.): "null".

Chapter One

Categories of Fractions

1. Categories of Fractions. Categories of Fractions and Adjoint Functors

1.1. A functor $F\colon \mathcal{C} \to \mathcal{D}$ is said make a morphism σ of $\mathcal{C}$ invertible if $F\sigma$ is invertible. We intend to associate with each category $\mathcal{C}$ and with each subset Σ of $\mathfrak{Ar}\,\mathcal{C}$ a category $\mathcal{C}[\Sigma^{-1}]$ and a functor $P_\Sigma\colon \mathcal{C} \to \mathcal{C}[\Sigma^{-1}]$ such that the following conditions are verified:

(i) P_Σ makes the morphisms of Σ invertible,

(ii) If a functor $F\colon \mathcal{C} \to \mathcal{X}$ makes the morphisms of Σ invertible, there exists one and only one functor $G\colon \mathcal{C}[\Sigma^{-1}] \to \mathcal{X}$ such that $F = G \cdot P_\Sigma$.

In order to do this, consider the diagram scheme T defined as follows: $\mathfrak{Ob}\, T$ coincides with $\mathfrak{Ob}\,\mathcal{C}$; $\mathfrak{Ar}\, T$ is the direct sum $\mathfrak{Ar}\,\mathcal{C} \amalg \Sigma$; if in_1 and in_2 are the canonical injections of $\mathfrak{Ar}\,\mathcal{C}$ and Σ into $\mathfrak{Ar}\,\mathcal{C} \amalg \Sigma$, then

$$\mathfrak{d}_T \circ \mathrm{in}_1 = \mathfrak{d}_{\mathcal{C}} \qquad \mathfrak{d}_T \circ \mathrm{in}_2 = \mathfrak{r}_{\mathcal{C}} | \Sigma,$$

$$\mathfrak{r}_T \circ \mathrm{in}_1 = \mathfrak{r}_{\mathcal{C}} \qquad \mathfrak{r}_T \circ \mathrm{in}_2 = \mathfrak{d}_{\mathcal{C}} | \Sigma.$$

Now let $\mathcal{C}[\Sigma^{-1}]$ be the quotient of the category $\mathcal{P}a\, T$ of paths of T by the following relations:

(a) $(\mathrm{in}_1\, v) \circ (\mathrm{in}_1\, u) = \mathrm{in}_1 (v \circ u)$ if $v \circ u$ is defined in $\mathcal{C}$.

(b) $\mathrm{in}_1 (\mathrm{Id}_{\mathcal{C}}\, a) = \mathrm{Id}_{\mathcal{P}a\, T}\, a$ for each object a of $\mathcal{C}$.

(c) $\mathrm{in}_2\, \sigma \circ \mathrm{in}_1 \sigma = \mathrm{Id}_{\mathcal{P}a\, T} (\mathfrak{d}_{\mathcal{C}}\, \sigma)$ and $\mathrm{in}_1\, \sigma \circ \mathrm{in}_2\, \sigma = \mathrm{Id}_{\mathcal{P}a\, T} (\mathfrak{r}_{\mathcal{C}}\, \sigma)$ if $\sigma \in \Sigma$.

Finally, let $P_\Sigma\colon \mathscr{C} \to \mathscr{C}[\Sigma^{-1}]$ be the functor which induces on $\mathfrak{Ob}\,\mathscr{C}$ the identity map of $\mathfrak{Ob}\,\mathscr{C}$ onto $\mathfrak{Ob}\,\mathscr{C}[\Sigma^{-1}]$, and on $\mathfrak{Ar}\,\mathscr{C}$ the composition of in_1 with the canonical maps of $\mathfrak{Ar}\,T$ into $\mathfrak{Ar}\,\mathscr{P}a\,T$ and of $\mathfrak{Ar}\,\mathscr{P}a\,T$ into $\mathfrak{Ar}\,\mathscr{C}[\Sigma^{-1}]$.

1.2. *Lemma: For each category* X, *the functor* $\mathscr{H}om(P_\Sigma, X)$: $\mathscr{H}om(\mathscr{C}[\Sigma^{-1}], \mathscr{X}) \to \mathscr{H}om(\mathscr{C}, \mathscr{X})$ *is an isomorphism from* $\mathscr{H}om(\mathscr{C}[\Sigma^{-1}], \mathscr{X})$ *onto the full subcategory of* $\mathscr{H}om(\mathscr{C}, \mathscr{X})$ *whose objects are the functors* F: $\mathscr{C} \to \mathscr{X}$ *which make all the morphisms of* Σ *invertible.*

The proof is left to the reader. This lemma states more precisely conditions (i) and (ii). From now on, we will say that $\mathscr{C}[\Sigma^{-1}]$ is *the category of fractions* of $\mathscr{C}$ for Σ, and that P_Σ is the canonical functor; we will say that the set of morphisms σ of $\mathscr{C}$ such that $P_\Sigma\,\sigma$ is invertible is the saturation of Σ.

1.3. *Proposition: Let the functor* $D\colon \mathscr{D} \to \mathscr{C}$ *be right-adjoint to* $G\colon \mathscr{C} \to \mathscr{D}$; *let* $\Phi\colon GD \to \mathrm{Id}\,\mathscr{D}$ *be an adjunction morphism from* D to G, *and* Σ *the set of morphisms* u *of* $\mathscr{C}$ *such that* $G\,u$ *is invertible. Then the following statements are equivalent:*

(i) D *is fully faithful.*

(ii) *The functor morphism* $\Phi\colon GD \to \mathrm{Id}\,\mathscr{D}$ *is invertible.*

(iii) *The functor* $H\colon \mathscr{C}[\Sigma^{-1}] \to \mathscr{D}$ *such that* $G = H \circ P_\Sigma$ *is an equivalence.*

(iv) *For each category* $\mathscr{X}$, *the functor* $\mathscr{H}om(G, \mathscr{X})$: $\mathscr{H}om(\mathscr{D}, \mathscr{X}) \to \mathscr{H}om(\mathscr{C}, \mathscr{X})$ *is fully faithful.*

(i)$\Leftrightarrow$(ii): For each morphism $\alpha\colon d \to d'$ of $\mathscr{D}$, we have $\alpha \circ (\Phi\,d) = (\Phi\,d') \circ (GD\,\alpha)$, which is equivalent to the Commutativity of the diagram of Fig. 1:

$$\begin{array}{ccc} \mathscr{D}(d, d') & \xrightarrow{\;D(d,d')\;} & \mathscr{C}(Dd, Dd') \\ \big\downarrow{\scriptstyle \mathscr{D}(\Phi d, d')} & & \big\downarrow{\scriptstyle G(Dd, Dd')} \\ \mathscr{D}(G\,Dd, d') & \xleftarrow{\;\mathscr{D}(G\,Dd,\, \Phi d)\;} & \mathscr{D}(G\,Dd, G\,Dd') \end{array}$$

Fig. 1

But the composition $\mathscr{D}(GD\,d, \Phi\,d') \circ G(Dd, Dd')$ is simply $\varphi(Dd, d')$, where φ is the adjunction isomorphism associated with Φ. It follows that $D(d, d')$ is a bijection for all d and d' if and only if $\mathrm{Hom}(\Phi\,d, d')$ is a bijection for all d and d', i.e. if and only if Φ is invertible.

(ii) $\Rightarrow$ (iii). By (ii), $H(P_\Sigma D)$ is isomorphic to $\mathrm{Id}\,\mathscr{D}$. Hence it is sufficient to prove that $(P_\Sigma D)H$ is isomorphic to $\mathrm{Id}\,\mathscr{C}[\Sigma^{-1}]$. Since $\mathscr{H}om(P_\Sigma, \mathscr{C}[\Sigma^{-1}])$ is fully faithful by lemma 1.2, it is sufficient to prove that $(P_\Sigma D)\,H\,P_\Sigma$ is isomorphic to P_Σ. But if $\Psi\colon \mathrm{Id}\,\mathscr{C} \to DG$, an adjunction morphism from G to D, is quasi-inverse to Φ, we know that the composition

$$G \xrightarrow{G\Psi} GDG \xrightarrow{\Phi G} G$$

is the identity of G. Since ΦG is invertible, $G\Psi$ is invertible; hence $P_\Sigma \Psi$: $P_\Sigma \to P_\Sigma DHP_\Sigma$ is an isomorphism by definition of Σ.

(iii) $\Rightarrow$ (iv). Since H is an equivalence, $\mathcal{H}om(H, \mathcal{X})$ is fully faithful. On the other hand, lemma 1.2 shows that $\mathcal{H}om(P_\Sigma, \mathcal{X})$ is fully faithful.

(iv) $\Rightarrow$ (ii). The proof is based on the following lemma:

1.3.1. *Lemma: Consider the following diagram of* $\mathcal{C}at$:

$$\begin{array}{ccc} \mathcal{C} & \xleftarrow{D} & \mathcal{D}' \\ & {\scriptstyle G}\searrow \quad \nearrow{\scriptstyle F} & \\ & \mathcal{D} & \end{array}$$

and let Φ: $(FG)D \to \operatorname{Id} \mathcal{D}'$ *be and adjunction morphism from* D *to* FG. *If for each category* $\mathcal{X}$ *,the functor*

$$\mathcal{H}om(G, \mathcal{X}): \mathcal{H}om(\mathcal{D}, X) \to \mathcal{H}om(\mathcal{C}, \mathcal{X})$$

is fully faithful then Φ *is an adjunction morphism from* GD *to* F.

Proof: Let Ψ: $\operatorname{Id}\mathcal{C} \to D(FG)$ be an adjunction morphism quasi inverse to Φ. By the hypothesis, there exists a unique functor morphism Ψ': $\operatorname{Id} \mathcal{D} \to GDF$, such that $\Psi' G = G\Psi$. Hence it is sufficient to prove that Ψ' and Φ are quasi-inverse to each other.

But the composition $D \xrightarrow{\Psi D} DFGD \xrightarrow{D\Phi} D$, after left-multiplication by G, gives $GD \xrightarrow{G\Psi D} GDFGD \xrightarrow{GD\Phi} GD$. Since $(D\Phi) \circ (\Psi D)$ is the identity of D, and since $G\Psi = \Psi' G$, it is clear that $(GD\Phi) \circ (\Psi' GD)$ is the identity of GD.

On the other hand, in order to prove that the composition

$$F \xrightarrow{F\Psi} FGDF \xrightarrow{\Phi F} F$$

is the identity of F, it is sufficient to prove that the composition

$$FG \xrightarrow{D\Psi' G} FGDFG \xrightarrow{\Phi FG} FG$$

is the identity of FG. This follows from the equation $\Psi' G = G\Psi$ and from the fact that Ψ is quasi-inverse to Φ.

Let us apply the lemma to the case where F: $\mathcal{D} \to \mathcal{D}'$ is the identity functor of $\mathcal{D}$ (hence $\mathcal{D} = \mathcal{D}'$). We see then that Φ: $GD \to \operatorname{Id} \mathcal{D}$ is an adjunction morphism from GD to $\operatorname{Id} D$, and hence is invertible. (We know that adjoint functors and adjunction morphisms are unique up to isomorphism; hence GD is isomorphic to $\operatorname{Id} \mathcal{D}$, and Φ is "isomorphic" to the functor morphism which is the identity of $\operatorname{Id} \mathcal{D}$.)

Note: The formulation of the dual to proposition 1.3 is left to the reader. In what follows we will refer to 1.3 either for the proposition itself or its dual.

1.4. Suppose that the equivalent conditions of 1.3 are verified, and let $\tau\colon T\to\mathscr{D}$ be a diagram of D. Then $\Phi\tau\colon GD\tau\to\tau$ is a diagram isomorphism so that $\varinjlim\tau$ exists if and only if $\varinjlim GD\tau$ exists. Since G is left adjoint to D, and hence commutes with direct limits, we see that the existence of $\varinjlim D\tau$ implies that of $\varinjlim\tau$ and of the following isomorphisms:

$$\varinjlim\tau \xrightarrow{\varinjlim(\Phi\tau)^{-1}} \varinjlim GD\tau \xrightarrow{\text{can}} G(\varinjlim D\tau).$$

Dually, D commutes with inverse limits. Moreover, *the existence of* $\varprojlim D\tau$ *implies that of* $\varprojlim\tau$: Let $L=\varprojlim D\tau$, then, since D is fully faithful it is sufficient to show that L is isomorphic to the image under D of an object of $\mathscr{D}$. If Ψ is an adjunction morphism quasi-inverse to Φ, it is then sufficient to prove that $\Psi L\colon L\to DGL$ is invertible. But for each object t of T, there exists a unique morphism $p_t\colon GL\to\tau t$ such that $pr_t=(Dp_t)\circ(\Psi L)$; in fact, with the usual notation, p_t is the image of pr_t under $\varphi(L,\tau t)$. By unicity of the morphisms p_t, (p_t) is a projective cone with domain vertex GL and base τ. Hence there exists a morphism $p\colon DGL\to L$ such that $pr_t\circ p=Dp_t$; hence $pr_t\circ p\circ(\Psi L)=pr_t$ for all t. It follows that $p\circ(\Psi L)=\operatorname{Id} L$; and if we set $L'=DGL$, and $i=\Psi L$, the equalities $(\Psi L')\circ i=(DGi)\circ(\Psi L)$ and $(\Psi L)\circ p=(DGp)\circ(\Psi L')$ imply that $p\circ(\Psi L')^{-1}\circ(DGi)$ is an inverse of ΨL. (The above argument is due to L. Gruson).

1.5. Some examples where proposition 1.3 can be applied:

1.5.1. Here, $\mathscr{C}$ is the category $\mathscr{T}\!op$ of topological spaces, $\mathscr{D}$ the category of Hausdorff topological spaces, D the inclusion functor. The functor G, left adjoint to D, associates to each topological space $\mathscr{X}$ the "largest Hausdorff quotient" of $\mathscr{X}$. Let $\tau\colon T\to\mathscr{D}$ be a small diagram: by 1.4, the inverse limits in the category of Hausdorff spaces coincide with the inverse limits in the category of all topological spaces; moreover $\varinjlim\tau$ exists and is identified with the largest Hausdorff quotient of $\varinjlim D\tau$.

1.5.2. Let $\mathscr{A}$ be an abelian category with enough injectives, and let $\mathscr{C}$ be the category $K^+(\mathscr{A})$ whose objects are the complexes

$$X_*\colon \cdots\to X_{n-1}\xrightarrow{d_{n-1}} X_n\xrightarrow{d_n} X_{n+1}\to\cdots$$

of $\mathscr{A}$ such that X_n is zero when n is near $-\infty$; a morphism of $K^+(\mathscr{A})$ is an equivalence class of homotopic chain homomorphisms (Verdier [*11*]). Now let $\mathscr{D}$ be the full subcategory of $K^+(\mathscr{A})$ formed by all complexes K_* such that X_n is injective for all n, and let D be the inclusions of $\mathscr{D}$ into $\mathscr{C}$. Then the functor G associates with each complex an injective resolution of this complex.

1.5.3. *Kelley Spaces:* A *Kelley space* is a Hausdorff topological space X such that a subset F of X is closed whenever its intersection with each compact subspace of X is closed. Each locally compact space is a Kelley space; each direct sum of Kelley spaces is a Kelley space; each closed subspace and each Hausdorff quotient of a Kelley space is a Kelley space.

We will write $\mathcal{Ke}$ for the full subcategory of $\mathcal{Top}$ formed by all Kelley spaces. If Y is a topological Hausdorff space, $Y_{\mathcal{Ke}}$ will denote the Kelley space which has the same underlying set as Y, and whose closed sets are the subsets F whose intersections with each compact subspace of Y is closed. The topology of $Y_{\mathcal{Ke}}$ is finer than that of Y; and each continuous map $f: X \to Y$ from a Kelley space X to Y can be factored through $Y_{\mathcal{Ke}}$. An equivalent statement of this property is that the functor $Y \rightsquigarrow Y_{\mathcal{Ke}}$ is right adjoint to the inclusion of $\mathcal{Ke}$ into the full subcategory of $\mathcal{Top}$ formed by all Hausdorff spaces.

It follows from 1.2 and 1.4 that the category $\mathcal{Ke}$ admits direct and inverse limits, and that we can describe them as follows: let $i: \mathcal{Ke} \to \mathcal{Top}$ be the inclusion, and $d: T \to \mathcal{Ke}$ a small diagram of range $\mathcal{Ke}$. Then $\varinjlim d$ is the largest Hausdorff quotient of $\varinjlim i \circ d$ (this quotient is a Kelley space). In particular, if $\varinjlim i \circ d$ is Hausdorff, $\varinjlim d$ coincides with $\varinjlim i \circ d$. Similarly, we have $\varprojlim d \simeq (\varprojlim i \circ d)_{\mathcal{Ke}}$; hence the underlying set of the inverse limit in $\mathcal{Ke}$ is the inverse limit of the underlying sets, but the topology is finer than the inverse limit topology. We will use the notations $(\varprojlim d)_{\mathcal{Ke}}$ or $(X \times Y)_{\mathcal{Ke}}$ instead of $\varprojlim d$ and $X \times Y$ to distinguish between inverse limits in $\mathcal{Ke}$ and in $\mathcal{Top}$.

Since a Kelley space X is the direct limit (in $\mathcal{Top}$ or in $\mathcal{Ke}$) of its compact subspaces K, each open subset U of X is the direct limit of the locally compact subspaces $K \cap U$. Hence it follows that *an open subset of a Kelley space is a Kelley space.*

1.5.4. *Groupoids.* If $\mathcal{C}$ is a category and if Σ is the set of all morphisms of $\mathcal{C}$, it is clear that the category $\mathcal{C}[\Sigma^{-1}]$ of fractions of $\mathcal{C}$ for Σ is a groupoid. When $\mathcal{C}$ in the category of paths of a diagram scheme T, we say that $C[\Sigma^{-1}]$ in *the groupoid of paths of* T. In general, we say that $\mathcal{C}[\Sigma^{-1}]$ is *the groupoid associated with* $\mathcal{C}$. It is clear that the functor $\mathcal{C} \rightsquigarrow \mathcal{C}[\Sigma^{-1}]$ is left adjoint to the inclusion of $\mathcal{Gr}$ in $\mathcal{Cat}$.

Similarly, for each category $\mathcal{C}$, $\mathcal{C}^*$ will be the subcategory of $\mathcal{C}$ which has the same objects as $\mathcal{C}$, and whose morphisms are the invertible morphisms of $\mathcal{C}$. It is clear that $\mathcal{C}^*$ is a groupoid and that the functor $\mathcal{C} \rightsquigarrow \mathcal{C}^*$ is right adjoint to the inclusion of $\mathcal{Gr}$ in $\mathcal{Cat}$.

Let us apply now the results of 1.3 and 1.4. Since the inclusion $i: \mathcal{Gr} \to \mathcal{Cat}$ is fully faithful and has a left adjoint, and since each small diagram of $\mathcal{Cat}$ has a direct limit, each small diagram of $\mathcal{Gr}$ will also have a direct limit. Moreover, i commutes with direct limits, since it has a right adjoint. In other words, if $d\colon T\to\mathscr{C}at$ is a small diagram and if $d(t)$ is a groupoid for each object t of T, then the direct limit of d in $\mathscr{C}at$ is a groupoid and is a direct limit of the diagram $T\to\mathscr{G}r$ induced by d.

Since the inclusion $i\colon\mathscr{G}r\to\mathscr{C}at$ has a right adjoint and is fully faithful and since each small diagram of $\mathscr{C}at$ has an inverse limit, this conclusion also holds for $\mathscr{G}r$. Moreover i commutes with inverse limits, since it has a left adjoint. Hence "the inverse limits in $\mathscr{G}r$ can be constructed like those of $\mathscr{C}at$".

2. The Calculus of Fractions

Let us return to the situation of proposition 1.3.

$$\mathscr{C}\underset{G}{\overset{D}{\leftrightarrows}}\mathscr{D}\qquad \Phi\colon GD\to \mathrm{Id}\,\mathscr{D}.$$

We will see first that we can give a simple construction of $\mathscr{C}[\Sigma^{-1}]$ by means of the functor DG and of a functor morphism $\Psi\colon \mathrm{Id}\,\mathscr{C}\to DG$ quasi-inverse to Φ. To that end, let us associate with each morphism $\gamma\colon c\to DGc'$ of $\mathscr{C}$ the diagram

$$\begin{array}{ccc} c & & c' \\ & {}_{\gamma}\searrow \quad \swarrow_{\Psi c'} & \\ & DGc' & \end{array}$$

Since Ψ is quasi inverse to Φ, $(\Phi G)\circ(G\Psi)$ is the identity of G, so that $G\Psi$, and hence $P_\Sigma\Psi$ is an isomorphism. It follows in particular that we have a map

$$\gamma \rightsquigarrow \gamma_* = (P_\Sigma \Psi c')^{-1}\circ(P_\Sigma\gamma)$$

from $\mathscr{C}(c, DGc')$ to $\mathscr{C}[\Sigma^{-1}](c, c')$.

Lemma: The map $\gamma\rightsquigarrow\gamma_*$ *is a bijection from* $\mathscr{C}(C, DGc')$ *onto* $\mathscr{C}[\Sigma^{-1}](c, c')$.

Proof: If we define H by the equation $G=H\circ P_\Sigma$, it follows that

$$H(c, c')\gamma_* = (HP_\Sigma\Psi c')^{-1}\circ(HP_\Sigma\gamma) = (G\Psi c')^{-1}\circ(G\gamma).$$

Since $(G\Psi c')^{-1} = \Phi Gc'$, we finally have

$$H(c, c')\gamma_* = \varphi(c, Gc')\gamma,$$

where $\varphi(c, Gc')\colon\mathscr{C}(c, DGc')\xrightarrow{\sim}\mathscr{D}(Gc, Gc')$ is the adjunction isomorphism associated with Φ. Since $H(c, c')$ is a bijection, this completes the proof.

Hence we can identify the set $\mathscr{C}[\Sigma^{-1}](c, c')$ with $\mathscr{C}(c, DGc')$. If $\gamma\colon c\to DGc'$ and $\gamma'\colon c'\to DGc''$ are two morphisms of $\mathscr{C}$, it is clear that the composition $\gamma'_* \circ \gamma_*$ is of the form γ''_*, where γ'' is the composition $(\Psi DGc'')^{-1} \circ (DG\gamma') \circ \gamma$ obtained from the diagram of Fig. 2:

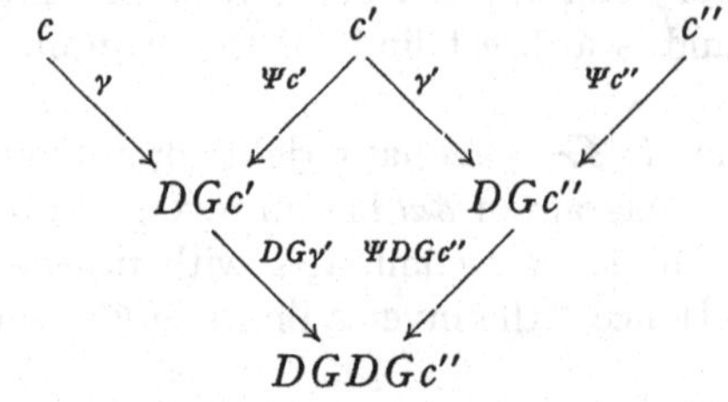

Fig. 2

2.2. More generally, this leads us to consider the subsets Σ of Ar $\mathscr{C}$ such that:

a) The identities of $\mathscr{C}$ are in Σ.

b) If $u\colon X \to Y$ and $v\colon Y \to Z$ are in Σ, their composition $v \circ u$ is also in Σ.

c) For each diagram $X' \xleftarrow{s} X \xrightarrow{u} Y$ where $s \in \Sigma$, there exists a commutative square

$$\begin{array}{ccc} X & \longrightarrow & Y \\ \downarrow & & \downarrow{\scriptstyle t} \\ X' & \overset{u'}{\dashrightarrow} & Y' \end{array} \quad \text{where} \quad t \in \Sigma.$$

d) If $f, g\colon X \rightrightarrows Y$ are morphisms of $\mathscr{C}$ and if $s\colon X' \to X$ is a morphism of Σ such that $fs = gs$, there exists a morphism $t\colon Y \dashrightarrow Y'$ of Σ such that $tf = tg$:

$$X' \longrightarrow X \underset{g}{\overset{f}{\rightrightarrows}} Y \overset{t}{\dashrightarrow} Y'.$$

2.3. When a), b), c), d) are satisfied, we say that Σ *admits a calculus of left fractions*. This terminology is justified by the following simple description of the morphisms of $\mathscr{C}[\Sigma^{-1}]$: for each object c of C, let $c\backslash\Sigma$ be the full subcategory of $c\backslash\mathscr{C}$ whose objects are the morphisms $s\colon c \to c'$ belonging to Σ. (Recall that the objects of the category $c\backslash\mathscr{C}$ of objects of $\mathscr{C}$ under c are the morphisms of $\mathscr{C}$ with domain c; if $s\colon c \to c'$ and $t\colon c \to c''$ are two such morphisms, a morphism from s to t is a morphism $\gamma\colon c' \to c''$ of $\mathscr{C}$ such that $\gamma \cdot s = t$.) If d is another object of $\mathscr{C}$, the direct limit $\varinjlim_s \mathscr{C}(d, \mathfrak{r}s)$ of the functor

$$s\colon c \to c' \rightsquigarrow \mathscr{C}(d, c'),$$

from $c\backslash\Sigma$ to $\mathscr{E}$ (as usual, $\mathscr{E}$ is the category of sets) can be described as follows:

Let $\boldsymbol{H}(d, c)$ be the set of pairs (s, f) of morphisms of $\mathscr{C}$ such that $\mathfrak{t}f = \mathfrak{t}s$, $\mathfrak{b}f = d$, $\mathfrak{b}s = c$ and $s \in \Sigma$.

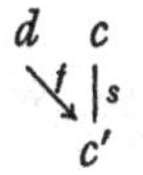

Two pairs (s, f) and (t, g) belonging to $\boldsymbol{H}(d, e)$ are said to be equivalent if there exists a commutative diagram (Fig. 3)

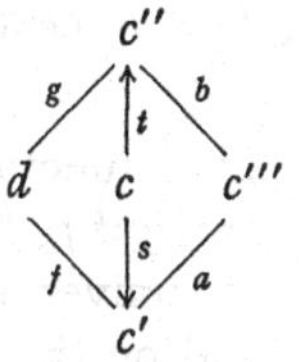

Fig. 3

such that $as = bt$ belongs to Σ (the commutativity of this diagram is equivalent to the equalities $as = bt$ and $af = bg$). It follows from a), b), c), d) that this defines an equivalence relation on $\boldsymbol{H}(d, c)$ and that $\varinjlim_s \mathscr{C}(d, \mathfrak{t}s)$ is identified with the quotient of $\boldsymbol{H}(d, c)$ by this relation.

Let $s|f$ be the canonical image of an element (s, f) of $\boldsymbol{H}(d, c)$ in $\varinjlim_s \mathscr{C}(d, \mathfrak{t}s)$. Define then a new category $\Sigma^{-1}\mathscr{C}$ as follows: the objects of $\Sigma^{-1}\mathscr{C}$ coincide with those of $\mathscr{C}$, and hence with those of $\mathscr{C}[\Sigma^{-1}]$; if d and c are two such objects, $\Sigma^{-1}\mathscr{C}(d, c)$ is equal to $\varinjlim_s \mathscr{C}(d, \mathfrak{t}s)$; finally the composition maps $\Sigma^{-1}\mathscr{C}(e, d) \times \Sigma^{-1}\mathscr{C}(d, c) \to \Sigma^{-1}\mathscr{C}(e, c)$ are defined by the formula:

$$(s|f) \circ (t|g) = s' s | f' g,$$

where s' is an element of Σ and f' a morphism which makes the diagram of Fig. 4 commutative:

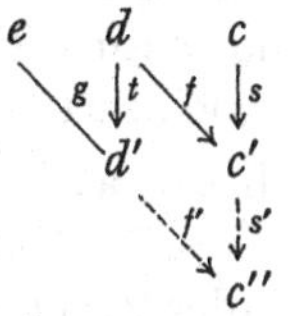

Fig. 4

Conditions a), b), c), d) show that the above formula makes sense and makes $\Sigma^{-1}\mathscr{C}$ a category. It is this category which we want to compare with $\mathscr{C}[\Sigma^{-1}]$. In order to do this, let d and e be two objects of $\mathscr{C}$, and s an object of $c\backslash\Sigma$. Passing to direct limits, the maps

$$(s, f) \rightsquigarrow (P_\Sigma s)^{-1} P_\Sigma f$$

induce a map

$$\pi(d, c): s f \rightsquigarrow (P_\Sigma s)^{-1} P_\Sigma f$$

from

$$\Sigma^{-1}\mathscr{C}(d, c) \quad \text{to} \quad \mathscr{C}[\Sigma^{-1}](d, c).$$

2.4. *Proposition: Let* $\mathscr{C}$ *be a category, and* Σ *a subset of* $\mathfrak{Ar}\, C$ *which admits a calculus of left fractions. The identity map of* $\mathfrak{Ob}\, C$ *and the maps* $\pi(d, c)$ *define an isomorphism from* $\Sigma^{-1}\mathscr{C}$ *onto* $\mathscr{C}[\Sigma^{-1}]$. *In particular the map* $\pi(d, c)$: $s|f \rightsquigarrow (P_\Sigma s)^{-1} \circ P_\Sigma f$ *is a bijection from* $\varinjlim_s \mathscr{C}(d, \mathfrak{t}s)$ *onto* $\mathscr{C}[\Sigma^{-1}](d, c)$.

Proof: Let ${}_\Sigma P: \mathscr{C} \to \Sigma^{-1}\mathscr{C}$ be the functor defined by the equations: ${}_\Sigma P c = c$ if $c \in \mathfrak{Ob}\,\mathscr{C}$ and ${}_\Sigma P f = (\mathrm{Id}\,\mathfrak{t}f|f)$ if $f \in \mathfrak{Ar}\,\mathscr{C}$. It is clear then that the pair $(\Sigma^{-1}G, {}_\Sigma P)$ is a solution of the universal problem 1.1: for ${}_\Sigma P$ makes the morphisms of Σ invertible, and on the other hand, if $F: \mathscr{G} \to \mathscr{X}$ is a functor which makes the morphisms of Σ invertible, then $F = G \circ {}_\Sigma P$ if and only if $G_{\mathfrak{Ob}} = F_{\mathfrak{Ob}}$ and $G(s|f) = F(s)^{-1} \circ F(f)$, whenever $f \in \mathfrak{Ar}\, C$, $s \in \Sigma$ and $\mathfrak{t}f = \mathfrak{t}s$.

In particular, if $\mathscr{X}$ is equal to $\mathscr{C}[\Sigma^{-1}]$ and F is equal to P_Σ, G is simply the functor π defined by the maps $\pi(d, c)$ of 2.3. Since $(\mathscr{C}[\Sigma^{-1}], P_\Sigma)$ and $(\Sigma^{-1}C, {}_\Sigma P)$ are solutions of the same universal problem, π is an isomorphism.

In what follows, we will identify $\Sigma^{-1}\mathscr{C}$ with $\mathscr{C}[\Sigma^{-1}]$ under π, whenever Σ admits a calculus of left fractions. We will also use freely the dual to proposition 2.4. In particular, we will say that a subset of $\mathfrak{Ar}\,\mathscr{C}$ *admits a calculus of right fractions* if the duals of conditions a), b), c), d) of 2.2 are satisfied.

2.5. Some examples of applications:

a) Let $G: \mathscr{C} \to \mathscr{D}$ be left adjoint to $D: \mathscr{D} \to \mathscr{C}$; suppose that D is fully faithful (see 1.3), and let $\Psi: \mathrm{Id}\,\mathscr{C} \to DG$ be an adjunction morphism from G to D. Let Σ be the intersection of all subsets of $\mathfrak{Ar}\,\mathscr{C}$ which contain the identity morphisms, the morphisms $\Psi c: c \to DGc$, and which are stable under composition (i.e. which satisfy b) above).

Then, by definition, Σ satisfies a) and b). In order to verify c), it is obviously sufficient to take $s: X \to X'$ of the form $\Psi X: X \to DGX$, and to chose $t = \Psi Y$, $u' = DGu$. In order to verify d), we may take $s: X' \to X$, of the form $\Psi c: c \to DGc$, and choose $t = \Psi Y: Y \to DGY$; we then have the commutative squares:

$$\begin{array}{ccc} DGc & \xrightarrow{f} & Y \\ {\scriptstyle \Psi DGc}\downarrow & & \downarrow{\scriptstyle \Psi Y} \\ DGDGc & \xrightarrow{DGf} & DGY \end{array} \quad \text{and} \quad \begin{array}{ccc} DGc & \xrightarrow{g} & Y \\ {\scriptstyle \Psi DGc}\downarrow & & \downarrow{\scriptstyle \Psi Y} \\ DGDGc & \xrightarrow{DGg} & DGY \end{array}$$

Moreover ΨDG and $DG\Psi$ have the same inverse functor isomorphism $D\Phi G$, so that they coincide. Hence we have the following equalities:

$$(DGf)\circ(\Psi DGc)=(DGf)\circ(DG\Psi c)=(DGg)\circ(DG\Psi c)=(DGg)\circ(\Psi DGc).$$

Hence $$(\Psi Y)\circ f=(\Psi Y)\circ g \quad \text{Q.E.D.}$$

b) Let $\mathscr{C}$, $\mathscr{D}$, G, D be as in a), but let Σ be the set of morphisms made invertible by G. As above, we can verify that Σ admits a calculus of left fraction.

c) Let $\mathscr{A}$ be an abelian category, and $K(\mathscr{A})$ the "category of complexes of $\mathscr{A}$ up to homotopy" (an object of $K(\mathscr{A})$ is a complex

$$X_* \cdots \to X_{n-1} \xrightarrow{d_{n-1}} X_n \xrightarrow{d_n} X_{n+1} \to \cdots$$

of $\mathscr{A}$; a morphism of $K(\mathscr{A})$ is an equivalence class of homotopic chain homomorphisms). Let Σ be the set of morphisms u: $X_* \to Y_*$ of $K(\mathscr{A})$ such that $H_n(u)$: $H_n(X_*) \to H_n(Y_*)$ is an isomorphism for each n. We can verify (see VERDIER) that Σ admits a calculus of left and right fractions (i.e. Σ satisfies conditions a), b), c), d) and their duals). The category $K(\mathscr{A})[\Sigma^{-1}]$ is called the "derived category" of $\mathscr{A}$.

d) Let $\mathscr{A}$ be an abelian category, and $\mathscr{B}$ a thick subcategory of $\mathscr{A}$; i.e. $\mathscr{B}$ is a full subcategory of $\mathscr{A}$, and for each exact sequence

$$0 \to A' \xrightarrow{u} A \xrightarrow{v} A'' \to 0$$

of $\mathscr{A}$, A is an object of $\mathscr{B}$ if and only if A' and A'' are objects of $\mathscr{B}$. If $\Sigma(\mathscr{B})=\Sigma$ denotes the set of morphisms s of $\mathscr{A}$ such that Ker s and Coker s are in $\mathscr{B}$, Σ admits a calculus of left and right fractions. The category $\mathscr{A}[\Sigma^{-1}]$ is then identified with quotient category $\mathscr{A}/\mathscr{B}$ (see GROTHENDIECK).

e) Let $\mathscr{A}$ be an abelian category and Σ the set of essential extensions of $\mathscr{A}$ (a monomorphism u: $A' \to A$ of $\mathscr{A}$ is an essential extension if, for each morphism v: $A \to A''$, the condition "vu=monomorphism" implies that "v=monomorphism"). We can verify that Σ admits a calculus of right fractions (but, in general, not of left fractions). The category $\mathscr{A}[\Sigma^{-1}]$ is the "spectral category of $\mathscr{A}$"; if $\mathscr{A}$ is the category of left unitary modules over a ring A with unit (more generally, if each sub-object of an object of $\mathscr{A}$ has a complement, in particular, if each object of $\mathscr{A}$ has an injective envelope), $A[\Sigma^{-1}]$ is an abelian category in which every morphism splits. If, moreover, A is left Noetherian (more generally, if $\mathscr{A}$ is locally Noetherian: see GABRIEL), each object of $\mathscr{A}[\Sigma^{-1}]$ is a direct sum of simple objects, and there is a bijection from the set of isomorphism classes of simple objects of $\mathscr{A}[\Sigma^{-1}]$ onto the set of isomorphism classes of indecomposable injectives of $\mathscr{A}$.

f) Let $\mathscr{C}$ be the category $\mathscr{E}$ of sets, and let σ be an infinite cardinal number. Let Σ_σ be the subset of $\mathfrak{Ar}\,\mathscr{E}$ formed by all morphisms $u: M \to N$ such that:

(i) $N \neq \emptyset \Rightarrow M \neq \emptyset$,

(ii) Card $(N - u(M)) < \sigma$,

(iii) There exists a subset M' of M such that the restriction $u|M'$ of u to M' injective and such that Card $(M - M') < \sigma$.

Consider also the subfamily $\Sigma_{\sigma*}$ of Ar $\mathscr{E}$ formed by all morphisms satisfying (ii) and (iii) only. Then $\Sigma_\sigma \subset \Sigma_{\sigma*}$; and Σ_σ and $\Sigma_{\sigma*}$ admit a calculus of left fractions.

3. Calculus of Left Fractions and Direct Limits

3.1. *Proposition: Let $\mathscr{C}$ be a category, and Σ a subset of $\mathfrak{Ar}\,\mathscr{C}$ which admits a calculus of left fractions. Then the canonical functor*

$$P_\Sigma: \mathscr{C} \to \mathscr{C}[\Sigma^{-1}]$$

commutes with finite direct limits.

Proof: Let T be a finite diagram scheme, $\tau: T \to \mathscr{C}$, a diagram of type T, in_t the canonical map of range $\varinjlim \tau$ whose domain is the image under τ of the object t of T. We want to show that, for every object c of $\mathscr{C}$, the morphisms $P_\Sigma(\mathrm{in}_t)$ induce a bijection

$$\mathscr{C}[\Sigma^{-1}]\left(P_\Sigma \varinjlim_t \tau t, c\right) \simeq \varprojlim_t \mathscr{C}[\Sigma^{-1}](P_\Sigma \tau t, c).$$

According to the description of the morphisms of $\mathscr{C}[\Sigma^{-1}]$ given in the preceding paragraph, we just have to prove that the canonical map

$$\varinjlim_{s \in c\backslash\Sigma} \mathscr{C}\left(\varinjlim_t \tau t, \mathfrak{r} s\right) \to \varprojlim_t \varinjlim_{s \in c\backslash\Sigma} \mathscr{C}(\tau t, \mathfrak{r} s)$$

or, equivalently, the canonical map

$$\varinjlim_s \varprojlim_t \mathscr{C}(\tau t, \mathfrak{r} s) \to \varprojlim_t \varinjlim_s \mathscr{C}(\tau t, \mathfrak{r} s)$$

is a bijection. This last result comes from the fact that in $\mathscr{E}$ finite inverse limits commute with quasi filtering direct limits.

3.2. *Corollary 1: Under the hypothesis of 3.1, if each finite diagram of $\mathscr{C}$ has a direct limit, the same holds for $\mathscr{C}[\Sigma^{-1}]$.*

It is sufficient to prove that each finite family of objects of $\mathscr{C}[\Sigma^{-1}]$ has a direct sum and that each pair of morphisms has a cokernel.

If $(c_t)_{t\in T}$ is a finite family of objects of $\mathscr{C}[\Sigma^{-1}]$ and if $\left(\coprod_{t\in T} c_t, \mathrm{in}_t\right)$ is a direct sum of this family in $\mathscr{C}$, it follows from the proposition that $\left(\coprod_{t\in T} c_t, P_\Sigma\, \mathrm{in}_t\right)$ is a direct sum of $(c_t)_{t\in T}$ in $\mathscr{C}[\Sigma^{-1}]$.

If $f, g\colon c \rightrightarrows d$ is a pair of morphisms of $\mathscr{C}[\Sigma^{-1}]$, there exists a morphism $s\colon d \to d'$ of Σ and two morphisms $f', g'\colon c \rightrightarrows d'$ of $\mathscr{C}$ such that $P_\Sigma f' = (P_\Sigma s) f$ and $P_\Sigma g' = (P_\Sigma s) g$. It follows from the preceding proposition that $(P_\Sigma e, P_\Sigma(\pi \circ s))$ is a cokernel of (f, g) in $\mathscr{C}[\Sigma^{-1}]$ if (e, π) is a cokernel of (f', g') in $\mathscr{C}$.

3.3. *Corollary 2: Under the hypothesis of 3.1, if $\mathscr{C}$ is an additive category, $\mathscr{C}[\Sigma^{-1}]$ is additive.*

Let o be a zero object of C. We will show first that o is also a zero object of $\mathscr{C}[\Sigma^{-1}]$. Since o is obviously an initial object of $\mathscr{C}[\Sigma^{-1}]$ (by the above proposition), let us prove that it is terminal. For this, it is sufficient to prove that, for each diagram of the form

$$\begin{array}{ccc} c & & o \\ & \searrow^{f} & \downarrow{\scriptstyle s} \\ & & d \end{array}$$

such that $s \in \Sigma$, we have $s|f = (\mathrm{Id}\, o)|0^c$, where $0^c \in C(c, o)$. Consider the diagram $o \xrightarrow{s} d \underset{\mathrm{Id}\, d}{\overset{s \cdot 0^d}{\rightrightarrows}} d$; there exists, according to condition d) of paragraph 2, an element $t\colon d \to d'$ of Σ such that $t = t(\mathrm{Id}\, d) = t s 0^d$. It follows that we have $tf = ts\, 0^d f = ts\, 0^c$, hence $s|f = (\mathrm{Id}\, o)|0^c$.

Now we note that the sets $\mathscr{C}[\Sigma^{-1}](c, d)$ can be identified with the direct limits $\varinjlim_{s \in d\backslash\Sigma} \mathscr{C}(c, \mathfrak{r}s)$, and hence, have a natural abelian group structure. Moreover, the composition laws are bilinear with respect to these group structures. The corollary follows from this fact, from the existence of a zero object and of finite sums (GABRIEL).

3.4. *Proposition: Let $G\colon \mathscr{C} \to \mathscr{D}$ be a functor commuting with finite direct limits, and Σ the subset of $\mathfrak{Ar}\,\mathscr{C}$ formed by all morphisms s such that Gs is an isomorphism. If finite direct limits exist in $\mathscr{C}$, Σ admits a calculus of left fractions; moreover, the functor $H\colon \mathscr{C}[\Sigma^{-1}] \to \mathscr{D}$ defined by the equation $G = H \circ P_\Sigma$ is conservative and commutes with finite direct limits.*

Conditions a) and b) are clearly verified. Let us check c): Since the square

$$\begin{array}{ccc} GX & \longrightarrow & GY \\ {\scriptstyle Gs}\downarrow & & \downarrow{\scriptstyle G(\mathrm{in}_2)} \\ GX' & \xrightarrow{G(\mathrm{in}_1)} & G\left(X \coprod_X Y\right) \end{array}$$

is cocartesian and since Gs is an isomorphism, $G(\mathrm{in}_2)$ is invertible, and hence in_2 belongs to Σ. For d): let $t\colon Y \to Y'$ be a cokernel of (f, g). Then

Gt is a cokernel of (Gf, Gg). But since Gs is invertible, we have $Gf = Gg$, and hence Gt is invertible and t belongs to Σ.

Let us show now that a morphism u of $\mathscr{C}[\Sigma^{-1}]$ is invertible if Hu is an isomorphism; we may indeed suppose that u is of the form $s|t$, where $s \in \Sigma$. Since $Hu = (Gs)^{-1} \circ (Gt)$ is an isomorphism, Gt is invertible, and hence t belongs to Σ, and the inverse of $s|t$ is $t|s$.

It remains to be shown that H commutes with finite direct limits: since each finite diagram of $\mathscr{C}[\Sigma^{-1}]$ has a direct limit (according to 3.2), it is sufficient to show that H commutes with finite sums and with cokernels of pairs of morphisms. The first statement is obvious. The second can be proved by an argument similar to that of corollary 1.

3.5. The above proposition can be applied in particular, to the case where T is a subset of $\mathfrak{Ar}\,\mathscr{C}$ admitting a calculus of left fractions, and where $G = P_T$. In that case, Σ is the saturation of T, and the categories $\mathscr{C}[\Sigma^{-1}]$ and $\mathscr{C}[T^{-1}]$ are identified. It is easily seen then that Σ is the subset of $\mathfrak{Ar}\,\mathscr{C}$ formed by the morphisms $u\colon c \to d$ which can be inserted in a commutative diagram of the form

$$\begin{array}{ccc} c & \xrightarrow{u} & d \\ {\scriptstyle s}\downarrow & \swarrow{\scriptstyle v} & \downarrow{\scriptstyle t} \\ c' & \xrightarrow{u'} & d' \end{array} \qquad \text{where} \quad s, t \in T.$$

Hence if $\mathscr{C}$ admits finite direct limits, the saturation of a subset of $\mathfrak{Ar}\,\mathscr{C}$ which admits a calculus of left fractions has the same property. We can then restrict ourselves to saturated subsets Σ of $\mathfrak{Ar}\,\mathscr{C}$ which admit a calculus of left fractions; and we will call *category of left fractions* of $\mathscr{C}$ a category of fractions with respect to such a subset Σ. We see that in a certain sense, the categories of left fractions of $\mathscr{C}$ classify the functors of domain $\mathscr{C}$ which commute with finite direct limits. We can show for instance that each category of left fractions of $\mathscr{E}$ is one of the categories $\mathscr{E}$, $\mathscr{E}[\Sigma_\sigma^{-1}]$ or $\mathscr{E}[\Sigma_{\sigma^*}^{-1}]$ for a suitable infinite cardinal number σ (see 2.5 f). Similarly, if we defined in a dual way the *categories of right fractions*, we would see that the only categories of right fractions of $\mathscr{E}$ are $\mathscr{E}$ and $\mathscr{E}[(\mathfrak{Ar}\,\mathscr{E})^{-1}]$.

3.6. Suppose now that $\mathscr{C}$ admits finite direct limits and finite inverse limits, and that Σ is a subset of $\mathfrak{Ar}\,\mathscr{C}$ admitting a calculus of left and right fractions. Then $\mathscr{C}[\Sigma^{-1}]$ admits finite direct and inverse limits and the functor $P_\Sigma\colon \mathscr{C} \to \mathscr{C}[\Sigma^{-1}]$ is exact. We can check moreover that a certain number of exactness properties of $\mathscr{C}$ are carried over to $\mathscr{C}[\Sigma^{-1}]$. We will restrict ourselves to the following property:

Proposition: If Σ admits a calculus of left and right fractions, and if $\mathscr{C}$ is an abelian category, $\mathscr{C}[\Sigma^{-1}]$ is abelian.

We know already that $\mathscr{C}[\Sigma^{-1}]$ is additive and admits kernels and cokernels. It remains to be shown that, for each morphism $u: c \to d$, the canonical morphism ϑ: Coim $u \to$ Im u is invertible. Since there exists a commutative diagram

$$\begin{array}{ccc} c & \xrightarrow{u} & d \\ & {\scriptstyle P_\Sigma u'} \searrow & \downarrow {\scriptstyle P_\Sigma s} \\ & & d' \end{array}$$

where $P_\Sigma s$ is invertible, we may suppose that u is of the form $P_\Sigma v$, where $v: c \to d$ is a morphism of $\mathscr{C}$. If η: Coim $v \to$ Im v is the canonical morphism defined by v in $\mathscr{C}$, we have $\vartheta = P_\Sigma \eta$. Hence ϑ is invertible if η is invertible.

Let us go back now to example d) of paragraph 2. We note (still under the hypothesis of the preceding proposition) that the full subcategory $\mathscr{C}(\Sigma)$ of $\mathscr{C}$ formed by the objects c of $\mathscr{C}$ such that $P_\Sigma c$ is zero, is thick (because P_Σ is exact). It follows easily that the map $\mathscr{B} \rightsquigarrow \Sigma(\mathscr{B})$ of paragraph 2 is a bijection from the set of thick subcategories of $\mathscr{C}$ onto the set of saturated subsets of $\mathfrak{Ar}\, C$ which admit a calculus of left and right fractions. Hence the notions of thick subcategories, of categories of left and right fractions, and of saturated subsets of $\mathfrak{Ar}\,\mathscr{C}$ admitting a calculus of left and right fractions, are in a one-to-one correspondance, if $\mathscr{C}$ is an abelian category.

4. Return to Paragraph 1

Let $\mathscr{C}$ be a category, and Σ a set of morphisms of $\mathscr{C}$ which admits a calculus of left fractions. We want to study the existence of a functor $D: \mathscr{C}[\Sigma^{-1}] \to \mathscr{C}$ right adjoint to P_Σ. We know that such a functor D exists if and only if, for each object e of $\mathscr{C}[\Sigma^{-1}]$, there exists an object d of $\mathscr{C}$ and a coliberty morphism $\gamma: P_\Sigma d \to e$.

4.1. *Definition: Let Σ be a set of morphisms of a category $\mathscr{C}$. We say that an object c of $\mathscr{C}$ is left closed for Σ if $\mathscr{C}(s, c)$ is a bijection for each morphism s of Σ.*

We have the following proposition:

Proposition: Let $\mathscr{C}$ be a category, and Σ a subset of $\mathfrak{Ar}\,\mathscr{C}$ admitting a calculus of left fractions. Then the following statements are equivalent:

(i) *P_Σ has a right adjoint (which is then fully faithful).*

(ii) *For each object c of $\mathscr{C}$, there exists an object d of $\mathscr{C}$ left-closed for Σ, and a morphism $s: c \to d$ such that $P_\Sigma s$ is invertible.*

Before giving a proof, we will give a few properties of left-closed objects for Σ, when Σ admits a calculus of left fractions.

4.1.1. *If Σ admits a calculus of left fractions, an object c of $\mathscr{C}$ is left closed for Σ if and only if $\mathscr{C}(s, c)$ is a surjection for each morphism s of Σ.*

It is sufficient to show that if $\mathscr{C}(s, c)$ is a surjection for each morphism s of Σ, then it is also an injection. Let

$$a \xrightarrow{s} b \overset{f}{\underset{g}{\rightrightarrows}} c, \quad s \in \Sigma$$

be such that $f \circ s = g \circ s$. According to 2.2.d), there exists a morphism $t: c \to c'$ such that $t \circ f = t \circ g$, with $t \in \Sigma$. Since $\mathscr{C}(t, c)$ is a surjection, there exists a morphism p such that $p \circ t = \mathrm{Id}\, c$. Hence the equation $t \circ f = t \circ g$ implies that $f = g$.

4.1.2. *If Σ admits a calculus of left fractions and if an object c of $\mathfrak{Ob}\,\mathscr{C}$ is left closed for Σ, the map $p_\Sigma(b, c): \mathscr{C}(b, c) \to \mathscr{C}[\Sigma^{-1}](b, c)$ is a bijection for each object b of $\mathfrak{Ob}\,\mathscr{C}$.*

We saw in 4.1.1 that for each morphism $s: c \to c'$ belonging to Σ, there exists a retraction $p: c' \to c$ of s. In particular, s is a monomorphism and the canonical map $\mathscr{C}(b, c) \to \varinjlim_s \mathscr{C}(b, c')$ is an injection. Since we have the diagram

$$c \xrightarrow{s} c' \overset{\mathrm{Id}\, c'}{\underset{sp}{\rightrightarrows}} c'$$

there exists a morphism $t: c' \to c''$ of Σ such that $t = tsp$. Hence for each morphism $f: b \to c'$, we have the following equalities: $(s|f) = (ts|tf) = (ts|tspf) = (\mathrm{Id}\, c|pf)$, so that the map $\mathscr{C}(b, c) \to \varinjlim_s \mathscr{C}(b, c')$ is a surjection. Then the statement follows immediately form 2.4.

4.2. *Proof of proposition* 4.1.

Consider two categories $\mathscr{C}$ and $\mathscr{D}$, functors $G: \mathscr{C} \to \mathscr{D}$ and $D: \mathscr{D} \to \mathscr{C}$, G being left adjoint to D, and a subset Σ of $\mathfrak{Ar}\,\mathscr{C}$ which admits a calculus of left fractions. If s belongs to Σ and d to $\mathfrak{Ob}\, D$, each adjunction isomorphism from G to D allows us to identify the map $\mathscr{C}(s, Dd)$ with $\mathscr{D}(Gs, d)$. It follows that if G makes the morphisms of Σ invertible, the objects Dd are left-closed for Σ.

In particular, if the functor $P_\Sigma: \mathscr{C} \to \mathscr{C}[\Sigma^{-1}]$ has a right adjoint i, $i P_\Sigma c$ is left-closed for Σ, for any $c \in \mathfrak{Ob}\,\mathscr{C}$. If Ψ is an adjunction morphism from P_Σ to i and Φ an adjunction morphism quasi inverse to Ψ, the morphism $\Psi c: c \to i P_\Sigma c$ is made invertible by P_Σ, since Φ is invertible [1.3. (ii)] and since $(\Phi P_\Sigma) \circ (P_\Sigma \Psi) = \mathrm{Id}\, P_\Sigma$. This proves that (i) implies (ii).

Conversely, suppose that with each object c of $\mathscr{C}$ we can associate a left-closed object $i(c)$ and a morphism $a(c): c \to i(c)$ such that $P_\Sigma a(c)$ is invertible. For each $d \in \mathfrak{Ob}\,\mathscr{C}$, $a(c)$ induces a bijection from $\mathscr{C}[\Sigma^{-1}](d, c)$ onto $\mathscr{C}[\Sigma^{-1}](d, i(c))$, this set being canonically isomorphic to $\mathscr{C}(d, i(c))$

by 4.1.2. Hence for each object c of $\mathscr{C}[\Sigma^{-1}]$, we get an isomorphism

$$\alpha(c): \mathscr{C}[\Sigma^{-1}](P_\Sigma d, c) \to C(d, i(c))$$

which is functorial in d. For each morphism $\gamma: c \to c'$ of $C[\Sigma^{-1}]$, there exists then a unique morphism $i(\gamma)$ such that $\mathscr{C}(d, i(\gamma)) \circ \alpha(c) = \alpha(c') \circ (\mathscr{C}[\Sigma^{-1}](P_\Sigma d, \gamma))$. The maps $c \rightsquigarrow i(c)$ and $\gamma \rightsquigarrow i(\gamma)$ define a functor i, right adjoint to P_Σ.

Chapter Two

Simplical Sets

1. Functor Categories

We wish first to complete some results of the dictionary. Notations will be the same.

1.1. Let $\mathscr{D}$ be a category, and $\mathscr{D}^\circ \mathscr{E}$ the category of contravariant functors from $\mathscr{D}$ to the category of sets $\mathscr{E}$. If d is an object of $\mathscr{D}$, it is clear that the functor

$$F \rightsquigarrow F(d)$$

from $\mathscr{D}^\circ \mathscr{E}$ to $\mathscr{E}$ commutes with direct and inverse limits. Moreover, a morphism f of $\mathscr{D}^\circ \mathscr{E}$ is an epimorphism (resp. monomorphism, resp. isomorphism) if and only if $f(d)$ is surjective (resp. injective, resp. bijective), for any $d \in \mathfrak{Ob}\, \mathscr{D}$.

We will see now that *each functor $F: \mathscr{D}^\circ \to \mathscr{E}$ is a direct limit of representable functors:* let $\mathscr{D}/F$ be the following category: the objects of $\mathscr{D}/F$ are the morphisms $\alpha: \mathscr{D}(a) \to F$ of $\mathscr{D}^\circ \mathscr{E}$; if $\alpha: \mathscr{D}(a) \to F$ and $\beta: \mathscr{D}(b) \to F$ are two such objects, $\mathrm{Hom}_{\mathscr{D}/F}(\alpha, \beta)$ is formed by all morphisms $f: a \to b$ of $\mathscr{D}$ such that $\alpha = \beta \circ \mathscr{D}(f)$; composition of morphisms in $\mathscr{D}/F$ is "obvious". There is then a "natural" morphism $d_F: \mathscr{D}/F \to \mathscr{D}^\circ \mathscr{E}$ which associates with an object $\alpha: \mathscr{D}(a) \to F$ the functor $\mathscr{D}(a)$; with a morphism $f: \alpha \to \beta$ of $\mathscr{D}/F$ the morphism $\mathscr{D}(f): \mathscr{D}(a) \to \mathscr{D}(b)$ of $\mathscr{D}^\circ \mathscr{E}$. Moreover, the morphisms $\alpha: \mathscr{D}(a) \to F$ induce a morphism $\varinjlim d_F \xrightarrow{\alpha F} F$ which is invertible, this being easily deduced from above.

Given two functors F and G from $\mathscr{D}^\circ$ to $\mathscr{E}$, each functor morphism $\varphi: F \to G$ induces a functor $\mathscr{D}/\varphi: \mathscr{D}/F \to \mathscr{D}/G$. This functor associates with an object $\alpha: \mathscr{D}(a) \to F$ of $\mathscr{D}/F$ the object $\varphi \circ \alpha: \mathscr{D}(a) \to G$ of $\mathscr{D}/G$; and its restriction to $\mathrm{Hom}_{\mathscr{D}/F}(\alpha, \beta)$ is the inclusion of $\mathrm{Hom}_{\mathscr{D}/F}(\alpha, \beta) \subset \mathrm{Hom}_{\mathscr{D}}(a, b)$ into $\mathrm{Hom}_{\mathscr{D}/G}(\varphi \circ \alpha, \varphi \circ \beta) \subset \mathrm{Hom}_{\mathscr{D}}(a, b)$. With these definitions, d_F is the composition

$$\mathscr{D}/F \xrightarrow{\mathscr{D}/\varphi} \mathscr{D}/G \xrightarrow{d_G} \mathscr{D}^\circ \mathscr{E}$$

and the canonical morphism $\varinjlim (d_G \circ (\mathscr{D}/\varphi)) \xrightarrow{l\varphi} \varinjlim d_G$ induces a commutative square

$$\begin{array}{ccc} \varinjlim d_F & \xrightarrow{l\varphi} & \varinjlim d_G \\ {\scriptstyle \alpha_F}\downarrow & & \downarrow{\scriptstyle \alpha_G} \\ F & \xrightarrow{\varphi} & G \end{array}$$

1.2. Recall that the usual construction of the direct limit $\varinjlim d_F$ is the following: let F_0 be the direct sum in $\mathscr{D}^\circ \mathscr{E}$ of the representable functors $\mathscr{D}(b)_\beta$, where β runs through the objects of $\mathscr{D}/F$, and where $\mathscr{D}(b)_\beta$ denotes the domain of the morphism β. Similarly, let F_1 be the direct sum of the representable functors $\mathscr{D}(a)_f$, where f runs through the morphisms $f\colon \alpha \to \beta$ of $\mathscr{D}/F$ and where $\mathscr{D}(a)_f$ denotes the domain of α. If in_f and in_β denote the canonical monomorphisms from $\mathscr{D}(a)_f$ and $\mathscr{D}(b)_\beta$ into F_1 and F_0, we define two functor morphisms $d_0, d_1\colon F_1 \rightrightarrows F_0$ by the relations $d_0 \circ \text{in}_f = \text{in}_\beta \circ \mathscr{D}(f)$ and $d_1 \circ \text{in}_f = \text{in}_\alpha$. Then $\varinjlim d_F$, and hence F, are identified with the cokernel of the pair d_0, d_1, which can be translated by saying that we have an *exact sequence*

$$F_1 \overset{d_1}{\underset{d_0}{\rightrightarrows}} F_0 \longrightarrow F$$

We will say that this exact sequence is the *canonical presentation* of F (compare with [Bourbaki, Algèbre commutative] 1.2.8); the functor F is in a certain sense an amalgamation of the representable functors $\mathscr{D}(b)_\beta$, the amalgamation conditions being given by the representable functors $\mathscr{D}(b)_f$ and the morphisms d_0 and d_1.

1.3. *Proposition: Let $\mathscr{C}$ be a category with direct limits, and $G\colon \mathscr{D}^\circ \mathscr{E} \to \mathscr{C}$, a functor. Then the following statements are equivalent:*

(i) *G commutes with direct limits.*

(ii) *G is left adjoint to a functor $D\colon \mathscr{C} \to \mathscr{D}^\circ \mathscr{E}$. Moreover, the functor $G \rightsquigarrow G \circ h^{\mathscr{D}}$ is an equivalence of the full subcategory of $\mathscr{H}om(\mathscr{D}^\circ \mathscr{E}, \mathscr{C})$ formed by the functors G which commute with direct limits on $\mathscr{H}om(\mathscr{D}, \mathscr{C})$.*

We know that each left adjoint functor commutes with direct limits. Conversely, suppose that G commutes with direct limits. We will then define a functor $D\colon \mathscr{C} \to \mathscr{D}^\circ \mathscr{E}$ right adjoint to G. Let D be the functor $c \rightsquigarrow h_c^{\mathscr{C}} \circ G \circ h^{\mathscr{D}}$, so that $D(c)$ is the functor $a \rightsquigarrow \mathscr{C}(G(h_a^{\mathscr{D}}), c)$. The adjunction isomorphism $\varphi(f, c)\colon \mathscr{C}(Gf, c) \simeq \mathscr{D}^\circ \mathscr{E}(f, Dc)$ can then be given explicitely as follows: if $\eta\colon Gf \to c$ is a morphism of $\mathscr{C}$ and a an object of $\mathscr{D}$, the image of η under $\varphi(f, c)$ maps an element $\lambda \in f(a)$ on the composition

$$G(h_a^{\mathscr{D}}) \xrightarrow{G(\alpha)} G(f) \xrightarrow{\eta} c$$

which is an element of $(Dc)(a)$ (α is the morphism associated with λ by the canonical bijection from $f(a)$ onto $\mathscr{D}^\circ\mathscr{E}(h_a^{\mathscr{D}}, f)$). If f is of the form $h_b^{\mathscr{D}}$, $\mathscr{C}(Gf, c)$ is simply $(Dc)(b)$, and it is easily seen that $\varphi(h_b^{\mathscr{D}}, c)$ is the canonical bijection from $(Dc)(b)$ onto $\mathscr{D}^\circ\mathscr{E}(h_b^{\mathscr{D}}, Dc)$. Hence $\varphi(f, c)$ is bijective when f is representable. Since each object of $\mathscr{D}^\circ\mathscr{E}$ is a direct limit of representable functors, since the functors $f \rightsquigarrow \mathscr{C}(Gf, c)$ and $f \rightsquigarrow \mathscr{D}^\circ\mathscr{E}(f, Dc)$ transform direct limits into inverse limits, and since $\varphi(f, c)$ is functorial in f, it follows that $\varphi(f, c)$ is bijective for all f.

It remains to be shown that the functor $G \rightsquigarrow G \circ h^{\mathscr{D}}$ is an equivalence, and hence, in particular, that each functor $G\colon \mathscr{D}^\circ\mathscr{E} \to \mathscr{C}$ which commutes with direct limits is determined by its restriction to representable functors. In order to do this, we associate with each functor $H\colon \mathscr{D} \to \mathscr{C}$ a functor $H'\colon \mathscr{D}^\circ\mathscr{E} \to \mathscr{C}$ defined as follows: If $F\colon \mathscr{D}^\circ \to \mathscr{E}$ is an object of $\mathscr{D}^\circ\mathscr{E}$, $H'F$ is defined to be the direct limit of the functor

$$H_F\colon \mathscr{D}(a) \xrightarrow{\alpha} F \rightsquigarrow Ha$$

from $\mathscr{D}/F$ to $\mathscr{C}$. If $\varphi\colon F \to G$ is a functor morphism, H_F is simply the composition $H_G \circ (D/\varphi)$, so that $H'\varphi$ can be chosen to be the canonical morphism $\varinjlim \mathscr{D}/\varphi\colon \varinjlim H_F \to \varinjlim H_G$. The reader will verify that $G \rightsquigarrow G \circ h^{\mathscr{D}}$ and $H \rightsquigarrow H'$ are quasi inverse to each other.

1.4. Here is an application of the preceding proposition: let g be a fixed object of $\mathscr{D}^\circ\mathscr{E}$. Since direct limits and products in $\mathscr{D}^\circ\mathscr{E}$ are computed "argument by argument", it is clear that the functor $?\times g\colon f \rightsquigarrow f \times g$ from $\mathscr{D}^\circ\mathscr{E}$ to $\mathscr{D}^\circ\mathscr{E}$ commutes with direct limits. Hence this functor is left adjoint to a functor from $\mathscr{D}^\circ\mathscr{E}$ to $\mathscr{D}^\circ\mathscr{E}$, which will be written $\mathscr{H}\!om_{\mathscr{D}^\circ\mathscr{E}}(g, ?)$. According to the construction used in the preceding proof, $\mathscr{H}\!om_{\mathscr{D}^\circ\mathscr{E}}(g, h)$ is, for each object h of $\mathscr{D}^\circ\mathscr{E}$, the functor $a \rightsquigarrow \mathscr{D}^\circ\mathscr{E}(h_a^{\mathscr{D}} \times g, h)$ from $\mathscr{D}^\circ$ to $\mathscr{E}$.

2. Definition of Simplicial Sets

2.1. Let Δ be the following category: the objects of Δ are the ordered sets $[n] = \{0, 1, 2, \dots n\}$, $n \in \mathbb{N}$, and its morphisms are non-decreasing maps. Consider in particular the following non decreasing maps:

— $\partial_n^i\colon [n-1] \to [n]$ is the increasing injection which does not take the value $i \in [n]$.

— $\sigma_n^i\colon [n+1] \to [n]$ is the non decreasing surjection which takes twice the value $i \in [n]$.

If there is no danger of confusion, we will write simply ∂^i and σ^i instead of ∂_n^i and σ_n^i.

These maps verify the following relations:

$$(*)\qquad \begin{aligned} &\partial^j_{n+1}\,\partial^i_n=\partial^i_{n+1}\,\partial^{j-1}_n && i<j\\ &\sigma^j_n\,\sigma^i_{n+1}=\sigma^i_n\,\sigma^{j+1}_{n+1} && i\leqq j\\ &\sigma^j_{n-1}\,\partial^i_n=\begin{cases}\partial^i_{n-1}\,\sigma^{j-1}_{n-2} & i<j\\ \mathrm{Id}\,[n-1] & i=j \text{ or } i=j+1\\ \partial^{i-1}_{n-1}\,\sigma^j_{n-2} & i>j+1.\end{cases} \end{aligned}$$

2.2. *Lemma: Every non decreasing map* $\mu\colon [m]\to[n]$ *can be written in one and only one way as*

$$(**)\qquad \mu=\partial^{i_s}_n\,\partial^{i_{s-1}}_{n-1}\cdots\partial^{i_1}_{n-t+1}\,\sigma^{j_t}_{m-t}\cdots\sigma^{j_2}_{m-2}\,\sigma^{j_1}_{m-1}$$

with $n\geqq i_s>\cdots>i_1\geqq 0$, $0\leqq j_t<\cdots<j_1<m$ *and* $n=m-t+s$.

Let $j_t<j_{t-1}<\cdots<j_1$ be the elements $j\in[m]$ for which $\mu(j)=\mu(j+1)$, and let $i_1<i_2<\cdots<i_s$ be the values of $i\in[n]$ which are not taken by μ. It is clear then that $(**)$ holds. Conversely, if μ can be written as in $(**)$, the indices i_k and j_e are easily characterized as above.

We will say that $(**)$ is the *canonical decomposition of* μ in Δ. It follows from the lemma that Δ *can be identified with the category* Δ' *generated by the objects* $[n]$, *the morphisms* ∂^i_n, σ^j_m, *and the relations* $(*)$: since the relations $(*)$ are verified in Δ, there exists a unique functor $\Delta'\to\Delta$ which induces the identity on the objects and on the morphisms ∂^i_n and σ^j_m. Since every morphism of Δ is a composition of morphisms ∂^i_n and σ^j_m, the map $\Delta'([m],[n])\to\Delta([m],[n])$ is a surjection. Let us show that it is an injection: let $\mu,\nu\colon[m]\to[n]$ be two morphisms of Δ' with the same image in $\Delta([m],[n])$. Since the relations $(*)$ are verified in Δ', it is easily seen that μ and ν have a decomposition in Δ' of the form $(**)$. (The unicity of such a decomposition is not given a priori). The images of these decompositions in Δ are the canonical decomposition of μ and ν. Since these images coincide, the decompositions of μ and ν will also coincide, and hence $\mu=\nu$. Q.E.D.

2.3. Finally, note that the epimorphisms $p\colon[m]\to[n]$ of Δ are simply the non decreasing surjections; such an epimorphism has a *section*, i.e. a morphism $s\colon[n]\to[m]$ such that $ps=\mathrm{Id}\,[n]$. Similarly, the monomorphisms $s\colon[n]\to[m]$ of Δ are the increasing injections; such a monomorphism has a *retraction* p, i.e. a morphism $p\colon[m]\to[n]$ such that $ps=\mathrm{Id}\,[n]$.

2.4. *Definition: Let* $\mathscr{C}$ *be a category. A simplicial object of* $\mathscr{C}$ is a *functor* $X\colon\Delta^\circ\to\mathscr{C}$ (resp. *a co-simplicial object of* $\mathscr{C}$ *is a functor* $Y\colon\Delta\to C$). *A morphism between simplicial* (resp. *co-simplicial*) *objects is a functor morphism.*

We will write X_n (resp. Y^n) for the image of the object $[n]$ of Δ° (resp. of Δ) under the functor X (resp. Y). Similarly, we will write ${}_X d_i^n$ and ${}_X s_j^m$, or simply d_i and s_j (resp. ${}^Y\partial_n^i$ and ${}^Y\sigma_m^j$ or simply ∂^i and σ^j) instead of $X(\partial_n^i)$ and $X(\sigma_m^j)$ (resp. $Y(\partial_n^i)$ and $Y(\sigma_m^j)$). We say that the d_i or the ∂^i are "face operators", the s_j or the σ^j "degeneracy operators". It is clear that the relations (*) are still verified when ∂_n^i and σ_m^j are replaced by ${}^Y\partial_n^i$ and ${}^Y\sigma_m^j$. Moreover, since Δ is defined by the objects $[n]$, the morphisms ∂_n^i and σ_m^j and the relations (*), the cosimplicial object Y is determined by the objects Y^n and the morphisms ∂^i and σ^j satisfying (*). If Y and Z are two cosimplicial objects of $\mathscr{C}$, a morphism $f\colon Y\to Z$ is a sequence of morphisms $f^n\colon Y^n\to Z^n$ of $\mathscr{C}$ such that

$$ {}^Z\partial_n^i\circ f^{n-1}=f^n\circ {}^Y\partial_n^i \quad\text{and}\quad {}^Z\sigma_n^i\circ f^{n+1}=f^n\circ {}^Y\sigma_n^i,\ \forall\, n, i. $$

Similarly, the morphisms d_i and s_j satisfy relations dual to (*):

$$ (***)\qquad \begin{aligned} & d_i^n\, d_j^{n+1}=d_{j-1}^n\, d_i^{n+1} && i<j\\ & s_i^{n+1}\, s_j^n=s_{j+1}^{n+1}\, s_i^n && i\leqq j\\ & d_i^n\, s_j^{n-1}=\begin{cases} s_{j-1}^{n-2}\, d_i^{n-1} & i<j\\ \mathrm{Id} & i=j \text{ or } i=j+1\\ s_j^{n-2}\, d_{i-1}^{n-1} & i=j+1. \end{cases} \end{aligned} $$

Moreover, the simplicial object X is determined by the objects X_n and the morphisms d_i and s_j satisfying (***). If X and T are two simplicial objects of $\mathscr{C}$, a morphism $g\colon X\to T$ is a sequence of morphisms $g_n\colon X_n\to T_n$ of $\mathscr{C}$ such that

$$ {}_T d_i^n\circ g_n=g_{n-1}\circ {}_X d_i^n \quad\text{and}\quad {}_T s_i^n\circ g_n=g_{n+1}\circ {}_X s_i^n,\ \forall\, i, n. $$

2.5. If $\mathscr{C}$ is the category $\mathscr{E}$ of sets, we say *simplicial set* or *complex* instead of simplicial object of $\mathscr{E}$. According to the notations of the Dictionary, $\Delta^\circ\mathscr{E}$ stands for the category of simplicial sets and $\Delta[n]$ is the complex $[p]\rightsquigarrow \Delta([p],[n])$. $\Delta[n]$ is called the *standard n-simplex.*

If X is an arbitrary complex, the elements of X_n are called *n-simplices* of X (if $n=0$, we say also a *vertex* of X; if $n=1$, we say also the *edges* of X). In particular, the identity map $\mathrm{Id}[n]$ is an n-simplex of $\Delta[n]$, called the fundamental simplex of $\Delta[n]$. We know (see Dictionary, under "representable") that for each x of X_n, there exists one and only one morphism from $\Delta[n]$ to X which sends the fundamental simplex of $\Delta[n]$ onto x. This morphism will be written

$$ (*)\qquad \tilde{x}\colon \Delta[n]\to X $$

and we will say that $\tilde{x}$ is the *singular n-simplex associated with x*, so that there is a canonical bijection between n-simplices and singular n-simplices.

If X is a complex, a *subcomplex* Y is by definition a subfunctor of X. This means that for each n, Y_n is contained in X_n, and that the face and degeneracy operators of Y are induced by those of X. Similarly, we call *quotient* of a complex X a complex Y such that for each n, Y_n is a quotient set of X_n, and such that the face and degeneracy operators of Y are induced by those of X.

Here are a few examples of simplicial sets.

2.5.1. *The complex* I_n: this is the direct limit of the diagram of Fig. 5

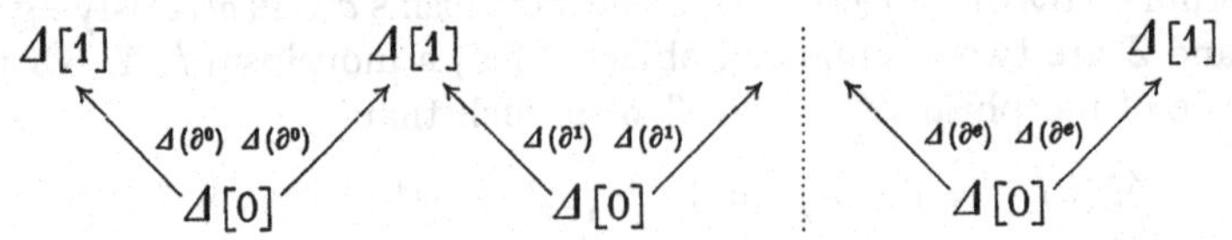

Fig. 5

where the standard simplex $\Delta[1]$ appears n times and where e is zero if n is even and 1 if n is odd.

We have in particular $I_1 = \Delta[1]$.

2.5.2. The *simplicial circle* Ω is the cokernel of the pair of morphisms $\Delta(\partial^0), \Delta(\partial^1): \Delta[0] \rightrightarrows \Delta[1]$.

2.5.3. If X and Y are two simplicial sets, we will write $\mathcal{H}om(X, Y)$ for the simplicial set $[n] \rightsquigarrow \Delta^\circ\mathcal{E}(\Delta[n]\times X, Y)$ of 1.4. Then we get a functor $\mathcal{H}om: (\Delta^\circ\mathcal{E})^\circ \times (\Delta^\circ\mathcal{E}) \to \Delta^\circ\mathcal{E}$; moreover, by 1.4. we have isomorphisms

$$\varphi: \Delta^\circ\mathcal{E}(X\times Y, Z) \simeq \Delta^\circ\mathcal{E}(X, \mathcal{H}om(Y, Z))$$

which are functorial in X, Y, Z.

3. Skeleton of a Simplicial Set

3.1. *Definition: Let X be a complex, and $x\in X_m$ an m-simplex of X. We say that x is degenerate if there exists an epimorphism $s: [m]\to[n]$ with $n<m$, and an n-simplex y such that $x = X(s)(y)$.*

Proposition (Eilenberg-Zilber lemma): *For each m-simplex x of X, there is an epimorphism $s: [m]\to[n]$ and a non-degenerate n-simplex y such that $x=X(s)(y)$. Moreover, the pair (s, y) is unique.*

Recall the proof given by Eilenberg and Zilber: the existence of s and y is obvious. Suppose that (s, y) and (s', y') are two pairs satisfying the above conditions, and let σ and σ' be sections of s and s'. We have then

$$x = X(s)(y) = X(s')(y'),$$

$$y = X(\sigma)(x), \quad y' = X(\sigma)(x).$$

Hence

$$y = X(\sigma)X(s')(y') = X(s'\sigma)(y').$$

Since y is non degenerate $s'\sigma$ is a monomorphism. If $[n]$ and $[n']$ are ranges of s and s', we have $n'\geqq n$, and hence, by "symmetry", $n'=n$. It follows that $s'\sigma$ is an increasing bijection, i.e. the identity map of $[n]$, and that $y=y'$. Moreover, we have seen that each section σ of s is a section of s', which implies $s=s'$.

3.2. We will give now another interpretation of Eilenberg-Zilber lemma: if $s\colon [m]\to[n]$ and $s'\colon [m]\to[n']$ are two *epimorphisms* of Δ, the diagram

$$\begin{array}{ccc} [m] & \xrightarrow{s} & [n] \\ {\scriptstyle s'}\downarrow & & \\ [n'] & & \end{array}$$

has clearly an amalgamated sum $[n']\overset{s',s}{\amalg}[n]$ in Δ. Then the functor $h^{\Delta}\colon \Delta\to\Delta^{\circ}\mathscr{E}$ commutes with this amalgamated sum, i.e. $\Delta\left([n']\overset{s',s}{\amalg}[n]\right)$ is identified with the amalgamated sum of the diagram

$$\begin{array}{ccc} \Delta[m] & \xrightarrow{\Delta(s)} & \Delta[n] \\ {\scriptstyle \Delta(s')}\downarrow & & \\ \Delta[n'] & & \end{array}$$

For let $\tilde{y}\colon \Delta[n]\to X$ and $\tilde{y}'\colon \Delta[n']\to X$ be two singular simplices such that $\tilde{y}\circ\Delta(s)=\tilde{y}'\circ\Delta(s')$. Let $y\in X_n$ and $y'\in X_{n'}$ be the simplices associated with $\tilde{y}$ and $\tilde{y}'$. We then have $X(s)(y)=X(s')(y')$. By the preceding lemma, there are two epimorphisms $t\colon [n]\to[p]$ and $t'\colon [n']\to[p']$ and non-degenerate simplices $z\in X_p$ and $z'\in X_{p'}$ such that $y=X(t)(z)$ and $y'=X(t')(z')$. Hence $X(ts)(z)=X(t's')(z')$ and $ts=t's'$, $z=z'$, by unicity. If $\tilde{z}$ is the singular simplex associated with z, we then have $\tilde{y}=\tilde{z}\circ\Delta(t)$ and $\tilde{y}'=\tilde{z}\circ\Delta(t')$. Moreover, the equality $ts=t's'$ implies the existence of a morphism $r\colon [n']\overset{s',s}{\amalg}[n]\to[p]$ such that $t=r\circ\mathrm{in}_2$ and $t'=r\circ\mathrm{in}_1$, where in_1 and in_2 are the canonical maps from $[n']$ and $[n]$ to $[n']\overset{s',s}{\amalg}[n]$. We then have $\tilde{y}=\tilde{z}(\Delta r)(\Delta\,\mathrm{in}_2)$ and $\tilde{y}'=\tilde{z}\Delta(r)\,\Delta(\mathrm{in}_1)$, so that $\tilde{y}$ and $\tilde{y}'$ can be factored through $\Delta\left([n']\overset{s',s}{\amalg}[n]\right)$. The unicity of this factorization follows, for instance, from the fact that in_1 is an epimorphism of Δ, and hence has a section; then $\Delta(\mathrm{in}_1)$ has a section and is an epimorphism of $\Delta^{\circ}\mathscr{E}$.

3.3. We will study now another application of Eilenberg-Zilber lemma: let $\underset{n}{\Delta}$ be the full subcategory of Δ formed by all objects $[p]$ such that $p\leqq n$. Each functor $Y\colon \underset{n}{\Delta}{}^{\circ}\to\mathscr{E}$ is determined by the sets $Y([p])$, which we will also write Y_p, and the maps $Y(\partial_j^i)$ and $Y(\sigma_{p-1}^j)$ which we will also write ${}_Yd_i^p$ and ${}_Ys_j^{p-1}$ or d_i and s_j $(p\leqq n)$. Such a functor is called a *truncated complex of order n*. Then the inclusion functor I_n:

$\underset{n}{\Delta}\to\Delta$ induces a restriction functor R_n: $\Delta^\circ\mathscr{E}\to\underset{n}{\Delta}{}^\circ E$ which can be given explicitely as follows: if X is a complex, $R_n X$ is the truncated complex Y defined by the equations

$$Y_p = X_p, \quad {}_Y d_i^p = {}_X d_i^p \quad \text{and} \quad {}_Y s_j^{p-1} = {}_X s_j^{p-1} \quad \text{for} \quad p \leqq n.$$

Since the p-simplices are identified with the singular p-simplices, we see that R_n is identified with the functor

$$X \rightsquigarrow \Delta^\circ\mathscr{E}(\Delta[?], X)$$

where ? takes all values $p\leqq n$. This functor is right adjoint to a functor G_n: $\underset{n}{\Delta}{}^\circ\mathscr{E}\to\Delta^\circ\mathscr{E}$ which commutes with direct limits and associates $\underset{n}{\Delta}[p]$ with $\Delta[p]$, for $p\leqq n$ (see 1.3). For each truncated complex Y, let us introduce (see 1.1 and 1.3) the following functors:

$$d_Y\colon \underset{n}{\Delta}[p] \xrightarrow{\beta} Y \rightsquigarrow \underset{n}{\Delta}[p]$$

$$d_Y^n\colon \underset{n}{\Delta}[p] \xrightarrow{\beta} Y \rightsquigarrow \Delta[p]$$

$$(d_Y^n)_q\colon \underset{n}{\Delta}[p] \xrightarrow{\beta} Y \rightsquigarrow \Delta[p]_q.$$

Then the morphism α_Y: $\varinjlim d_Y \to Y$ of 1.1. is invertible, $G_n(Y)$ is equal to $\varinjlim d_Y^n$ and $G_n(Y)_q$ to $\varinjlim (d_Y^n)_q$. But if q is smaller than n, we have $\varinjlim (d_Y^n)_q = (\varinjlim d_Y^n)_q$; thus $R_n G_n Y$ is simply $\varinjlim d_Y$, and we see easily that the adjunction morphism $\Psi_n(Y)$: $Y\to R_n G_n Y$ can be chosen equal to α_Y^{-1}. Consequently *Ψ_n is invertible and G_n is fully faithful* (I., 1.3).

3.4. We will look now at the adjunction morphism Φ_n: $G_n R_n \to \mathrm{Id}(\Delta^\circ\mathscr{E})$ quasi inverse to Ψ_n:

Proposition: Φ_n is a monomorphism, i.e. for each complex X and each integer $p\in\mathbb{N}$, $(\Phi_n X)_p$: $(G_n R_n X)_p \to X_p$ is an injection.

Let $Z = G_n R_n X$. We know that $R_n \Phi_n(X)$ is a retraction of $\Psi_n R_n(X)$. Since Ψ_n is invertible, $R_n \Phi_n$ is invertible and $(\Phi_n X)_p$ is a bijection for $p\leqq n$. On the other hand, we have seen that Z is the direct limit of the functor

$$\underset{n}{\Delta}[p] \xrightarrow{\beta} R_n X \rightsquigarrow \Delta[p]$$

from $\underset{n}{\Delta}/R_n X$ to $\Delta^\circ\mathscr{E}$. In particular, Z is the quotient of a direct sum of standard simplices $\Delta[p]$ with $p\leqq n$. But all q-simplices of such a $\Delta[p]$ are degenerate for $q>n$, so that the same thing holds for the q-simplices of Z. It is then sufficient to prove the following: *let ψ: $Z\to X$ be a morphism of complexes such that ψ_p is injective for $p\leqq n$ and that all the q-simplices of Z are degenerate for $q>n$. Then ψ_p is an injection for all p.*

Let z and z' be two q-simplices of Z such that $q>n$. By Eilenberg-Zilber lemma, there are epimorphisms $s\colon [q]\to[p]$ and $s'\colon [q]\to[p']$ and non degenerate simplices x and x' such that $z=Z(s)(x)$ and $z'=Z(s')(x')$; a fortiori, we have $p, p'\leqq n$. Since ψ_r is injective for $r\leqq n$, $\psi_p(x)$ and $\psi_{p'}(x')$ are non degenerate. Hence $(s, \psi_p(x))$ and $(s', \psi_{p'}(x'))$ are "Eilenberg-Zilber decompositions" of $\psi_q(z)$ and $\psi_q(z')$. The equality $\psi_q(z)=\psi_q(z')$ implies then $s=s'$, $p=p'$ and $x'=x'$. Hence $z=z'$. Q.E.D.

3.5. The *n-skeleton* Sk^nX *of a complex* X is the subcomplex Y of X such that Y_p is formed by all simplices degenerated from q-simplices such that $q\leqq n$ (i.e. Y_p is formed by all $x\in X_p$ such that there exists an epimorphism $s\colon [p]\to[q]$, $q\leqq n$, and a q-simplex y of X such that $x=X(s)(y)$).

Corollary 1: For each complex X, $\Phi_n(X)\colon G_nR_nX\to X$ induces an isomorphism from G_nR_nX into the n-skeleton Sk^nX of X.

We know already that $(\Phi_n(X))_p$ is an injection for each p. On the other hand, each simplex of G_nR_nX is degenerated from a p-simplex such that $p\leqq n$. The same holds for the image of G_nR_nX, which is than contained in Sk^nX. Moreover, if $s\colon [p]\to[q]$ is an epimorphism of Δ such that $q\leqq n$ and if y is a q-simplex of X, y is the image of some $z\in(G_nR_nX)_q$, since $(\Phi_n(X))_q$ is a bijection for $q\leqq n$. Hence $X(s)(y)$ is the image of a simplex degenerated from z, which proves that the map $(G_nR_nX)_p\to(Sk^nX)_p$ is a surjection.

3.6. From now on, we will say that *a complex X is of dimension $\leqq n$* if it coincides with its n-skeleton. Then X is isomorphic to G_nR_nX, and hence to the quotient of a direct sum of standard simplices $\Delta[p]$ such that $p\leqq n$. The converse is also true. Thus we have the following corollary:

Corollary 2: The functors G_n and R_n induce an equivalence between the category of truncated complexes of order n and the full subcategory of $\Delta^\circ\mathscr{E}$ formed by all complexes of dimension $\leqq n$.

Let Y be a subcomplex of X, y a p-simplex of Y, and (s, z) the "Eilbenberg-Zilber" decomposition of y in X. If σ is a section of s, we clearly have $z=X(\sigma)(y)$, so that z belongs to Y. It follows in particular that Y is of dimension $\leqq n$ if X is. In this case, the map $Y\rightsquigarrow R_nY$ is a bijection from the subcomplexes of X onto the subcomplexes of R_nX. Take for example $X=\Delta[n]$. It is clear then that the fundamental simplex of $\Delta[n]$ is the only non degenerate simplex of dimension $\geqq n$. Since, on the other hand, the fundamental simplex "generates" $\Delta[n]$, we see that $\Delta[n]$ is of dimension n, and that each subcomplex of $\Delta[n]$ different from $\Delta[n]$ is contained in the $(n-1)$-skeleton $Sk^{n-1}\Delta[n]$, which is usually written $\dot{\Delta}[n]$ (it is the "*boundary*" *of the standard simplex of dimension n*).

3.7. Consider now a quotient Y of the complex X. Then Y_p is the quotient of X_p by an equivalence relation R_p which satisfies the following condition: if $f: [p] \to [q]$ is a morphism of Δ, the relation $x \underset{R_q}{\sim} y$ implies

$$X(f)(x) \underset{R_p}{\sim} X(f)(y).$$

If X is of dimension $\leqq n$, the same holds for Y. The functor $Y \rightsquigarrow R_n Y$ defines then a bijection from the set of quotients of X onto the set of quotients of $R_n X$. Take for example *a complex of finite type* X: such a complex is, by definition, isomorphic to the quotient of the direct sum of a *finite* family of standard simplices. This simply means that X is of *finite dimension* and that, for each p, X_p has a *finite number of elements*. Hence, if n is large enough, the quotients of X are in a one-to-one correspondance with the quotients of $R_n X$. But the $(R_n X)_p$ are finite, and are in a finite number. Consequently, *the number of quotients of* $R_n X$, *and hence of* X, *is finite*. Similarly, we see that *the subcomplexes of* X *are in a finite number.*

3.8. Again, suppose that X is an arbitrary complex. Then X is the union of its skeletons:

$$\emptyset = Sk^{-1}X \subset Sk^0 X \subset Sk^1 X \subset \ldots \subset Sk^n X \subset \ldots.$$

The skeleton $Sk^0 X$ is of *dimension o (or discrete):* it has the same vertices as X and is isomorphic to the direct sum of a family of standard simplices $\Delta[0]$. More generally, given $Sk^{n-1}X$, we want to find $Sk^n X$. Let Σ^n be the set of non-degenerate n-simplices of X; let $(\Delta[n]_\sigma)_{\sigma \in \Sigma^n}$ be a family of standard n-simplices indexed by Σ^n; and let $\tilde{\sigma}: \Delta[n]_\sigma \to X$ be the singular simplex associated with the non degenerate simplex σ. We then have the following proposition:

Proposition: With the above notations, the square of Fig. 6.

$$\begin{array}{ccc} \coprod_{\sigma \in \Sigma^n} \dot{\Delta}[n]_\sigma & \longrightarrow & Sk^{n-1}X \\ \text{inclusion} \downarrow & & \downarrow \text{inclusion} \\ \coprod_{\sigma \in \Sigma^n} \Delta[n]_\sigma & \xrightarrow{(\tilde{\sigma})} & Sk^n X \end{array}$$

Fig. 6

is cocartesian for $n \geqq 0$.

By corollary 2 to 3.6. it is sufficient to check that for each $p \leqq n$, the square of Fig. 7

$$\begin{array}{ccc} \coprod_{\sigma \in \Sigma^n} (\dot{\Delta}[n]_\sigma)_p & \longrightarrow & (Sk^{n-1}X)_p \\ \text{inclusion} \downarrow & & \downarrow \text{inclusion} \\ \coprod_{\sigma \in \Sigma^n} (\Delta[n]_\sigma)_p & \xrightarrow{(\tilde{\sigma})_p} & (Sk^n X)_p \end{array}$$

Fig. 7

is cocartesian. This is clear when $p<n$, since, in that case, the vertical arrows are bijections. For $p=n$, the complement of $\coprod(\dot{\Delta}[n]_\sigma)_n$ in $\coprod(\Delta[n]_\sigma)_n$ is formed by the fundamental simplices $\mathrm{Id}[n]_\sigma$. Similarly, the complement of $(Sk^{n-1}X)_n$ in $(Sk^n X)_n$ is formed by all non degenerate n-simplices of X. Hence the map $(\tilde{\sigma})_n$ induces a bijection from the complement of $\coprod(\dot{\Delta}[n]_\sigma)_n$ onto the complement of $(Sk^{n-1}X)_n$, so that the square (**) is also cocartesian for $p=n$.

In other words, the above proposition means that $Sk^n X$ is obtained by attaching to $Sk^{n-1}X$ a certain number of standard simplices $\Delta[n]$, along the boundary $\dot{\Delta}[n]$ of these simplices.

3.9. To complete this paragraph, let us determine a useful presentation of $\dot{\Delta}[n]$. In order to do this, let us consider the diagram

$$(*)\qquad \coprod_{0\leq i<j\leq n}\Delta[n-2]_{i,j}\overset{u}{\underset{v}{\rightrightarrows}}\coprod_{0\leq i\leq n}\Delta[n-1]_i\xrightarrow{p}\Delta[n]$$

where $\Delta[n-1]_i$ and $\Delta[n-2]_{i,j}$ are copies of $\Delta[n-1]$ and $\Delta[n-2]$, where p is defined by the face operators $\Delta(\partial_n^i):\Delta[n-1]_i\to\Delta[n]$, and where u (resp. v) is induced by the morphisms

$$\Delta(\partial_{n-1}^{j-1}):\Delta[n-2]_{i,j}\to\Delta[n-1]_i\quad(\text{resp. }\Delta(\partial_{n-1}^{i}):\Delta[n-2]_{i,j}\to\Delta[n-1]_j).$$

We can verify then that, for each q, the image of the map

$$p_q:\coprod(\Delta[n-1]_i)_q\to\Delta[n]_q,$$

induced by p on the q-simplices, is the set of q-simplices of $\dot{\Delta}[n]$. More precisely, p_q induces a bijection from coker (u_q, v_q) onto $\dot{\Delta}[n]_q$. Hence p induces an isomorphism from coker (u, v) onto $\dot{\Delta}[n]$.

4. Simplicial Sets and Category of Categories

4.1. We know that the functor $\mathscr{C}\rightsquigarrow\mathfrak{Ob}\,\mathscr{C}$ from $\mathscr{Cat}$ to $\mathscr{E}$ commutes with inverse limits. This result can also be obtained by noting that this functor is right-adjoint to the functor Dis: $\mathscr{E}\to\mathscr{Cat}$, which associates with each set A the category Dis A, whose objects are the elements of A, and whose only morphisms are the identities. Such a category will be called a *discrete* category.

Now let $\mathscr{Or}$ be the category of ordered sets (with non-decreasing maps as morphisms). Let i: $\mathscr{Or}\to\mathscr{Cat}$ be the functor which associates with an ordered set A the following category iA: $\mathfrak{Ob}\,iA=A$, and $\mathfrak{Ar}\,iA$ is the subset of $A\times A$ formed by all pairs (a, b) such that $b\leq a$. The range and domain maps are the restrictions of the projections pr_1 and pr_2 of $A\times A$ onto A; finally, composition is defined by the formula: $(a,b)\circ(b,c)=(a,c)$.

Then, according to 1.3, the restriction H of i to Δ defines two functors

$$G\colon \Delta^\circ\mathscr{E} \to \mathscr{C}at \quad \text{and} \quad D\colon \mathscr{C}at \to \Delta^\circ\mathscr{E}$$

such that D is right adjoint to G and that $G \circ h^\Delta$ is isomorphic to H.

4.2. We want to describe more precisely the functor G:

Lemma: If $n > 2$, the inclusion of $\dot{\Delta}[n]$ into $\Delta[n]$ induces an isomorphism from $G\dot{\Delta}[n]$ onto $G\,\Delta[n] \simeq i[n]$. If $n=1$, $G\dot{\Delta}[n]$ is a discrete category; if $n=2$, $G\dot{\Delta}[n]$ has 3 objects, denoted by 0, 1 and 2; in addition to the identity morphisms, $G\dot{\Delta}[2]$ has 4 morphisms: $u\colon 1 \to 2$, $v\colon 0 \to 2$, $w\colon 0 \to 1$ and $u \circ w$.

The lemma is trivial if $n=1$. Suppose then that $n>1$. By 3.9, since G commutes with direct limits, $G\dot{\Delta}[n]$ is identified with the cokernel of the pair.

$$\coprod_{0 \leqq i < j \leqq n} G\Delta[n-2]_{i,j} \underset{Gv}{\overset{Gu}{\rightrightarrows}} \coprod_{0 \leqq i \leqq n} G\Delta[n-1]_i .$$

Write $[p_k]$ for the subset of $[p]$ formed by all integers different from k, and $[p_{k,l}]$ for the subset of $[p]$ formed by all integers different from k, l. With these notations, (Gu, Gv) is identified with the pair of morphisms.

$$\coprod_{0 \leqq i < j \leqq n} i[n_{i,j}] \underset{v'}{\overset{u'}{\rightrightarrows}} \coprod_{0 \leqq i \leqq n} i[n_i],$$

where u' and v' are induced by the inclusions of $[n_{i,j}]$ into $[n_i]$ and $[n_j]$ respectively. We construct the direct limit of this diagram of categories as explained in the dictionary. With the notations of this dictionary, X is identified with the diagram-scheme subordinated to $i[n]$. In particular, we can then deduce the structure of $G\dot{\Delta}[2]$. Similarly, when $n>2$, let

$$\alpha = (a, a_1) \circ (a_1, a_2) \circ \cdots \circ (a_k, b)$$

be a morphism of $\mathscr{P}a\, i[n]$. Then a, a_1, a_2 belong to the same $[n_i]$, since $n>2$. Hence (a, a_2) and $(a, a_1) \circ (a_1, a_2)$ have the same image in coker(u', v'), and the same holds for α and $(a, a_2) \circ \cdots \circ (a_k, b)$. By induction on k, we see then that α and (a, b) have the same image in coker (u', v'), which shows that two objects of coker (u', v') are connected by one and only one morphism.

Now let X be a simplicial set. We know that the square of Fig. 8

$$(*) \qquad \begin{array}{ccc} \coprod_{\sigma \in \Sigma^n} \dot{\Delta}[n]_\sigma & \longrightarrow & Sk^{n-1}X \\ \text{inclusion}\downarrow & & \downarrow\text{inclusion} \\ \coprod_{\sigma \in \Sigma^n} \Delta[n]_\sigma & \longrightarrow & Sk^n X \end{array}$$

Fig. 8

is cocartesian (see II.3) The same holds for the square of Fig. 9

$$(**)\qquad \begin{array}{ccc} \coprod_{\sigma\in\Sigma^n} G\dot{\Delta}[n]_\sigma & \longrightarrow & G\,Sk^{n-1}X \\ \downarrow & & \downarrow \\ \coprod_{\sigma\in\Sigma^n} G\Delta[n]_\sigma & \longrightarrow & G\,Sk^n X \end{array}$$

Fig. 9

since G commutes with direct limits. But, for $n>2$, the left arrow is an isomorphism by the lemma, and hence the same holds for the right one. Since $X=\varinjlim Sk^n X$, we have

$$G(X)=\varinjlim G(Sk^n X)=G(Sk^2 X).$$

But a description of $G(Sk^2X)$ can be given quite easily:

a) Clearly, $G\,Sk^0X=\operatorname{Dis} X_0$. (2.1)

b) Let us apply (**) to the case $n=1$. The objects of the category $G(Sk^1X)$ are the elements of X_0. With each element $x\in X_1$, we associate the morphism $x\colon d_1 x\to d_0 x$. Then the set of morphisms of $G(Sk^1X)$ is generated by the elements of X_1, these morphisms verifying the relation $s_0\,x_0=\operatorname{Id} x_0$, for each $x_0\in X_0$.

c) Finally, let us apply (**) to the case $n=2$. We have then the following proposition:

Proposition: Let X be a simplicial set and $\mathfrak{X}$ the *diagram scheme defined by the equations:* $\mathfrak{Ob}\,\mathfrak{X}=X_0$, $\mathfrak{Ar}\,\mathfrak{X}=X_1$, $\mathfrak{d}_{\mathfrak{X}}={}_Xd_1^1$ *and* $\mathfrak{r}_{\mathfrak{X}}={}_Xd_0^1$. *The category GX is then the quotient of the category* $\mathcal{P}a\,\mathfrak{X}$ *of paths of* $\mathfrak{X}$ *by the relations:*

$$s_0\,x=\operatorname{Id} x \quad \textit{if} \quad x\in X_0$$

and

$$(d_0\,\sigma)\circ(d_2\,\sigma)=d_1\,\sigma \quad \textit{if} \quad \sigma\in X_2.$$

4.3. *Corollary: Every adjunction morphism* $\Phi\colon GD\to\operatorname{Id}(\mathcal{C}at)$ *is an isomorphism (and hence, D is fully faithful).*

Recall that for each category $\mathcal{C}$, $(D\mathcal{C})_n=\mathcal{C}at(i[n],\mathcal{C})$. This implies that $(D\mathcal{C})_0=\mathfrak{Ob}\,\mathcal{C}$ and $(D\mathcal{C})_1=\mathfrak{Ar}\,\mathcal{C}$. Moreover, $(D\mathcal{C})_2$ is the set of triples (u,v,w) of $\mathfrak{Ar}\,\mathcal{C}$ such that $w=v\circ u$. From these equalities and from proposition 4.2 we get an isomorphism from $GD\mathcal{C}$ onto $\mathcal{C}$. This isomorphism is functorial in $\mathcal{C}$ and defines an adjunction morphism from D to G. Since an adjunction morphism is invertible, each adjunction morphism is invertible (I, 1.3).

5. Ordered Sets and Simplicial Sets: Shuffles

5.1. In 4.1, we defined a functor $i\colon \mathcal{O}r\to\mathcal{C}at$. This functor is fully faithful and it is right adjoint to the functor $O\colon \mathcal{C}at\to\mathcal{O}r$ defined as follows: for each category $\mathcal{C}$, we give to the set $\mathfrak{Ob}\,\mathcal{C}$ a preorder structure such that $a \leqq b$ if and only if there is a morphism of $\mathscr{C}$ with domain a and range b. Then $O\mathscr{C}$ is by definition the ordered set associated with the preordered set $\mathfrak{Ob}\,\mathscr{C}$.

Now let $Di = C$ and $OG = O^1$ (see 4.1). Then $C: \mathcal{O}r \to \Delta^\circ\mathscr{E}$ is right adjoint to $O^1: \Delta^\circ\mathscr{E} \to \mathcal{O}r$. Moreover, since both i and D are fully faithful, C is fully faithful. We can also verify directly that O^1C is isomorphic to $\mathrm{Id}\,\mathcal{O}r$, by writing the functor O^1 with the help of proposition 4.2: if X is a simplical set, we give to X_0 a preorder such that $x \leqq y$ if and only if there is a sequence $\sigma_1, \ldots, \sigma_n$ of elements of X, such that $d_1\sigma_1 = x$, $d_0\sigma_1 = d_1\sigma_2, \ldots, d_0\sigma_{n-1} = d_1\sigma_n$, $d_0\sigma_n = y$. O^1X is then the ordered set associated with this preordered set.

5.2. If E is an ordered set, CE is, by definition, the complex $[n] \rightsquigarrow \mathscr{C}at(i[n], iE)$. Since i is fully faithful, CE is then identified with the complex $[n] \rightsquigarrow \mathcal{O}r([n], E)$. In particular, $C[n]$ is identified with $\Delta[n]$. We will call i-th vertex of $\Delta[n]$ the image of the vertex of $\Delta[0]$ under the morphism η_i, which is associated with the map $0 \rightsquigarrow i$ from $[0]$ into $[n]$.

Similarly, let $\{n\}$ be the set $\{0, 1, \ldots, n\}$ ordered by the inequalities

$$0 < 1 > 2 < 3 > 4 \ldots$$

$C\{n\}$ is then simply I_n (2.5.1). If $\varepsilon_i: \Delta[0] \to I_n$ is the morphism associated with the map $0 \rightsquigarrow i$ from $[0]$ into $\{n\}$, we will write i for the i-th vertex of I_n, i.e. the image under ε_i of the vertex of $\Delta[0]$.

5.3. We will study now more closely the complex CE associated with an ordered set E: an n-simplex of CE is a non-decreasing map $x: [n] \to E$. Such a map defines a morphism $Cx: C[n] \to CE$ which is simply the singular n-simplex $\tilde{x}$ associated with x. The n-simplex x is non-degenerate if and only if x is an injection. Since C commutes with inverse limits, C transforms a monomorphism into a monomorphism. Hence, *for each nondegenerate simplex x of CE, the associated singular n-simplex $\tilde{x}: \Delta[n] \to CE$ is a monomorphism.* This implies that the image of a standard simplex in CE is always isomorphic to a standard simplex; for instance, CE contains no subcomplex isomorphic to the cokernel of the pair of morphisms $\Delta(\partial_1^0), \Delta(\partial_1^1): \Delta[0] \rightrightarrows \Delta[1]$.

5.4. *If E is a finite ordered set, CE is a complex of finite type*; for let a *chain* of E by any totally ordered finite subset of E; with each chain c containing $n_c + 1$ elements, we associate one and only one increasing map $x_c: [n_c] \to E$ whose image is c. If $c(1), \ldots, c(n)$ are the maximal chains of E, we then have a commutative diagram

$$(*) \qquad \coprod_{1 \leqq i < j \leqq n} [n_{c(i) \cap c(j)}] \underset{u}{\overset{v}{\rightrightarrows}} \coprod_{1 \leqq i \leqq n} [n_{c(i)}] \to E$$

where u (resp. v) is defined by the inclusion of $c(i) \cap c(j)$ into $c(i)$ [resp. into $c(j)$]. Moreover, for each integer p, the sequence

$$\coprod_{i<j} \mathcal{O}r([p], [n_{c(i) \cap c(j)}]) \rightrightarrows \coprod_{i} \mathcal{O}r([p], [n_{c(i)}]) \to \mathcal{O}r([p], E)$$

defined by $(*)$ is exact. This means that the image

$$\coprod_{i<j} \Delta[n_{c(i) \cap c(j)}] \rightrightarrows \coprod_{i} \Delta[n_{c(i)}] \xrightarrow{(\tilde{x}_{c(i)})} CE$$

of $(*)$ under the functor C is an exact sequence of $\Delta^\circ \mathcal{E}$ (we will say also that it is a "finite presentation" of CE).

5.5. This presentation will be used mainly when E is the ordered set $[p] \times [q]$. Since the functor C commutes with products, $C([p] \times [q])$ is identified with $\Delta[p] \times \Delta[q]$. Moreover, the number of maximal chains of $[p] \times [q]$ is $\binom{p+q}{p}$ and each of them has $p+q+1$ elements (see the Fig. 10, where $p=3$, $q=2$. See also MACLANE's "shuffles", [2] p. 243).

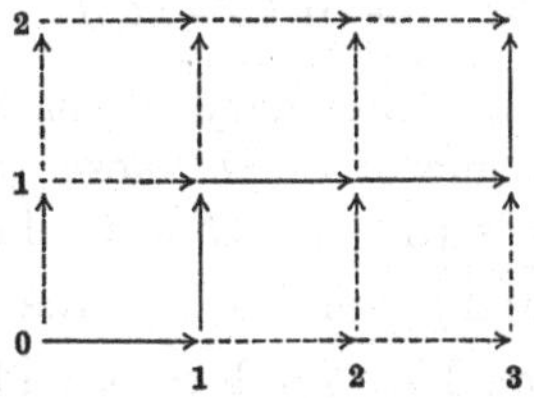

Fig. 10

We get then a finite presentation of $\Delta[p] \times \Delta[q]$

$$\coprod_{1 \leq i < j \leq \binom{p+q}{p}} \Delta[n_{c(i) \cap c(j)}] \rightrightarrows \coprod_{1 \leq i \leq \binom{p+q}{p}} \Delta[n_{c(i)}] \to \Delta[p] \times \Delta[q]$$

where $n_{c(i)} = p+q$ for each i.

6. Groupoids

Before studying the category $\mathcal{G}r$ in a similar way as above, we will establish some simple properties of this category.

6.1. *Examples.*

6.1.1. If E is a set, Dis E is a groupoid (4.1).

6.1.2. We know that the functor $G \rightsquigarrow \mathfrak{Ob}\, G$ from $\mathcal{G}r$ to $\mathcal{E}$ commutes with direct limits. We can see this if we note that this functor is left adjoint to a functor $Sc\colon \mathcal{E} \to \mathcal{G}r$: for each set E, the set of objects of $Sc\,E$ is E, the set of morphisms is $E \times E$, the domain and range-maps are the projections $\mathrm{pr}_2\colon (x, y) \rightsquigarrow y$ and $\mathrm{pr}_1\colon (x, y) \rightsquigarrow x$. Finally, composition in $Sc\,E$ is defined by the formula

$$(x, y) \circ (y, z) = (x, z).$$

Each groupoid G isomorphic to $Sc(\mathfrak{Ob}\, G)$ will be said *simply connected*. The functor $Sc\colon \mathscr{E} \to \mathscr{Gr}$ obviously commutes with inverse limits, but it does not commute with direct limits: indeed, using the notations of 2.1, the cokernel of $\partial^0, \partial^1\colon [0] \rightrightarrows [1]$ in $\mathscr{E}$ is the set $[0]$, while the cokernel of $Sc\,\partial^0, Sc\,\partial^1\colon Sc[0] \rightrightarrows Sc[1]$ is a groupoid with a single object whose automorphism group is $\mathbb{Z}$. However, we will use the following exactness property of the functor Sc: Let A and B be two sets, and let $C = A \cap B$, $D = A \cup B$. Then the inclusions define a commutative square

$$\begin{array}{ccc} Sc\,C & \to & Sc\,A \\ \downarrow & & \downarrow \\ Sc\,B & \to & Sc\,D \end{array}$$

which is cocartesian when C is not empty:

Let us show on this example how we can use the construction of direct limits given in the dictionary: for $d\colon T \to \mathscr{Cat}$ take the diagram

$$Sc\,B \leftarrow Sc\,C \to Sc\,A$$

keeping the notations of the dictionary. In particular, X is identified with the diagram sub-scheme of $Sc\,D$ whose set of objects is D and whose set of morphisms is $(A \times A) \cup (B \times B)$. Let $p\colon \mathscr{Pa}\,X \to \varinjlim d$ be the canonical projection and $j\colon \varinjlim d \to Sc\,D$ the functor defined by the above commutative square. Then j induces the identity on objects, and $j \circ p$, and hence also j, obviously induce a surjection on morphisms. In order to prove that j is an isomorphism, it remains to be shown that for two morphisms α and β of $\mathscr{Pa}\,X$, the equality $j\,p\,\alpha = j\,p\,\beta$ implies $\alpha \underset{S}{\sim} \beta$ (S is the relation defined in the dictionary).

Let c be an element of C. If (z, t) is a morphism of X, the relation $(z, t) \underset{S}{\sim} (z, c) \circ (c, t)$ holds in $\mathscr{Pa}\,X$. If $\alpha = (y, x_n) \circ \cdots \circ (x_1, x)$ is any morphism of $\mathscr{Pa}\,X$ with domain x and range y, we then have

$$\begin{aligned} \alpha &\underset{S}{\sim} (y, x_n) \circ \cdots \circ (x_1, c) \circ (c, x) \\ &\underset{S}{\sim} (y, x_n) \circ \cdots \circ (x_2, c) \circ (c, x_1) \circ (x_1, c) \circ (c, x) \\ &\underset{S}{\sim} (y, x_n) \circ \cdots \circ (x_2, c) \circ (c, x) \underset{S}{\sim} \cdots \underset{S}{\sim} (y, c) \circ (c, x). \end{aligned}$$

Hence each morphism of $\mathscr{Pa}\,X$ is equivalent to a morphism determined by its domain and its range. We then have $\alpha \underset{S}{\sim} \beta$ if α and β have the same domain and the same range, in particular if $j\,p\,\alpha = j\,p\,\beta$.

6.1.3. A groupoid is called *pointlike* if it has a single object. Each inverse and direct limit of pointlike groupoids is pointlike. Moreover it is clear that the full subcategory of Gr formed by all pointlike groupoids is equivalent to the category of groups. A groupoid G is called *connected* if $\mathfrak{Ob}\, G$ is not empty and if $G(x, y)$ is not empty for any pair of objects

(x, y). We will see in 6.1.5. that it is equivalent to say that the category G is equivalent to a pointlike groupoid.

6.1.4. Let G and H be groupoids. We say that H is a *subgroupoid* of G if H is a subcategory of G, i.e. if $\mathfrak{Ob}\, H$ and $\mathfrak{Ar}\, H$ are subsets of $\mathfrak{Ob}\, G$ and $\mathfrak{Ar}\, G$ and if the composition in H is induced by the composition in G. A maximal connected subgroupoid of G is called a *connected component* of G. It is clear that each groupoid G is isomorphic to the direct sum of its connected components, and we say that G is *totally disconnected* when all the connected components are pointlike.

6.1.5. *Each groupoid G is equivalent to a totally disconnected groupoid:* for let us choose an object x_i in each connected component i of G, and let $G((x_i))$ be the full subcategory of G whose objects are the x_i. We will see then that the inclusion j of $G((x_i))$ into G is an equivalence by giving an exhaustive construction of the functors $p\colon G \to G((x_i))$ such that $j \circ p$ is isomorphic to the identity functor of G and that $p \circ j$ is the identity functor of $G((x_i))$:

In order to do this, let us call a *tree* of G any subgroupoid which is isomorphic to a direct sum of simply connected groupoids (6.1.2). It is clear that each tree is contained in a maximal tree and that each maximal tree A of G has the same objects as G. Moreover, if a is an object of G contained in the connected component $i(a)$, A contains one and only one morphism $\varphi_a\colon x_{i(a)} \to a$. This allows us to define a functor $j_A\colon G \to G((x_i))$ which sends each object a onto $x_{i(a)}$ and each morphism $f\colon a \to b$ onto $\varphi_b^{-1} \circ f \circ \varphi_a$. This functor j_A satisfies the required conditions.

In the case where G has only one connected component containing the object x_0, the functor j_A makes the following square commutative:

$$\begin{array}{ccc} A & \xrightarrow{\text{inclusion}} & G \\ {\scriptstyle K}\downarrow & & \downarrow{\scriptstyle j_A} \\ \{x_0\} & \xrightarrow{\text{inclusion}} & G(x_0) \end{array}$$

where $\{x_0\}$ is a groupoid with a single morphism and a single object x_0. The reader can verify that *this square is cocartesian.*

6.2. Let G be a groupoid and g be an object of G. We will call *Poincaré group of G at g* (notation: $\Pi_1(G, g)$) the group $G(g, g)$ of automorphisms of g. If $\varphi\colon G \to H$ is a morphism of $\mathcal{G}r$, we will write $\Pi_1(\varphi, g)$ for the group-homomorphism.

$$\varphi(g, g)\colon G(g, g) \to H(\varphi g, \varphi g)$$

induced by φ.

Lemma ("Van Kampen"): *Let X, Y and Z be connected groupoids and*

$$Y \xleftarrow{\beta} Z \xrightarrow{\alpha} X$$

a diagram of $\mathcal{G}r$ such that α and β induce injections of $\mathfrak{Ob}\, Z$ into $\mathfrak{Ob}\, X$ and $\mathfrak{Ob}\, Y$. If z is an object of Z whose images in X, Y and $X \overset{Z}{\amalg} Y$ are respectively x, y and t, we have a canonical isomorphism

$$\Pi_1(X, x) \overset{\Pi_1(Z, z)}{\amalg} \Pi_1(Y, y) \simeq \Pi_1\big(X \overset{Z}{\amalg} Y, t\big)$$

where the first term is the amalgamated sum of the diagram of groups

$$\Pi_1(Y, y) \xleftarrow{\Pi_1(\beta, z)} \Pi_1(Z, z) \xrightarrow{\Pi_1(\alpha, z)} \Pi_1(X, x).$$

Let C be a maximal tree of Z, A and B maximal trees of X and Y which "extend" C. We then have a commutative diagram (Fig. 11)

$$(*) \qquad \begin{array}{ccccc} Y & \overset{\beta}{\leftarrow} & Z & \overset{\alpha}{\rightarrow} & X \\ | & & | & & | \\ B & \overset{\beta^1}{\leftarrow} & C & \overset{\alpha^1}{\rightarrow} & A \\ \downarrow & & \downarrow & & \downarrow \\ y & \leftarrow & z & \rightarrow & x \end{array}$$

Fig. 11

where α^1 and β^1 are induced by α and β, the other morphisms being the obvious ones. We know that there are two different ways of computing the direct limit of such a diagram (interversion of direct limits):

According to 6.1.5, the direct limits of the three columns of $(*)$ are canonically isomorphic to the pointlike groupoids $Y(y)$, $Z(z)$ and $X(x)$. Hence, passing to direct limits, the diagram $(*)$ induces morphisms

$$(**) \qquad Y(y) \overset{\beta_1}{\longleftarrow} Z(z) \overset{\alpha_1}{\longrightarrow} X(x)$$

obtained by restrictions of α and β. The direct limit of $(*)$ is then canonically isomorphic to that of $(**)$, which is a pointlike category whose Poincaré group is the amalgamated sum of the groups $\Pi_1(X, x)$ and $\Pi_1(Y, y)$ under $\Pi_1(Z, z)$.

On the other hand, the direct limits of the three rows of $(*)$ are identified respectively with $T = X \overset{Z}{\amalg} Y$, with a maximal tree D of T (see 6.1.2), and with $\{t\}$. Hence the direct limit of $(*)$ is isomorphic to the amalgamated sum of T and t under D, i.e. to $T(t)$ (6.1.5). The lemma follows from the comparison of the two preceding descriptions of the direct limit of the diagram $(*)$.

7. Groupoids and Simplicial Sets

7.1. Let j be the inclusion of $\mathcal{G}r$ into $\mathcal{C}at$ and Gr: $\mathcal{C}at \to \mathcal{G}r$ the functors $\mathscr{C} \rightsquigarrow \mathscr{C}[(\mathfrak{Ar}\,\mathscr{C})^{-1}]$ of I, 1.5.4. If G and D are the functors defined in § 4, let $\Pi = \mathrm{Gr} \circ G$ and $D^1 = D \circ j$. Then Π: $\Delta^\circ\mathscr{E} \to \mathcal{G}r$ is left to D^1:

$\mathcal{G}r \to \Delta^\circ\mathcal{E}$ and is determined up to isomorphism by $\Pi \circ h^\Delta$: according to 4.1, $G\Delta[n]$ is the category associated with the ordered set $[n]$; hence $\Pi\Delta[n]$ is the simply connected groupoid $Sc[n]$ (6.1.2). Finally, according to 4.3, D^1 *is fully faithful.*

If X is a simplicial set, proposition 4.2 allows us to give an explicit description of ΠX as follows:

The objects of the groupoid ΠX are the vertices of X, the morphisms of ΠX are generated by the elements of X_1 and their formal inverses; if $s \in X_1$, we have $\mathfrak{d}_{\Pi X}\, s = d_1 s$ and $\mathfrak{r}_{\Pi X}\, s = d_0 s$; the relations between the generating morphisms are generated by the following relations:

$$s_0 x = \mathrm{Id}\, x \quad \textit{if} \quad x \in X_0$$

and

$$(d_0 \sigma) \circ (d_2 \sigma) = d_1 \sigma \quad \textit{if} \quad \sigma \in X_2.$$

7.2. Let I_n be the simplicial set defined in 2.5.1, and in_i the canonical injection of the i-th copy of $\Delta[1]$ into I_n. Consider a morphism $f\colon I_n \to X$ and let x_i be the 1-simplex of X associated with the singular simplex $f \circ \mathrm{in}_i$. We define then a surjection from $\bigcup_{n \geq 0} \Delta^\circ\mathcal{E}(I_n, X)$ onto $\mathfrak{Ar}\, \Pi X$ by sending f onto the morphism $x_n^{(-1)^{n-1}} \circ \cdots \circ x_2^{-1} \circ x_1$ of ΠX. We could also restrict ourselves to the case where n is even. In that case, we could multiply an element of $\Delta^\circ\mathcal{E}(I_{2n}, X)$ by an element of $\Delta^\circ\mathcal{E}(I_{2p}, X)$, by "sticking I_{2n} and I_{2p} one at the end of the other, to obtain $I_{2(n+p)}$". The above surjection is then compatible with composition of morphisms.

7.3. For each simplicial set X, the groupoid ΠX is called the *Poincaré groupoid* of X. We will say that X is *connected* (resp. *simply connected*) if ΠX is connected (resp. simply connected). A connected component of ΠX with be called a *connected component of* X.

Thus two vertices x and y of X belong to a same connected component if and only if there is a morphism f from I_n into X such that $f(0) = x$, $f(n) = y$ (see 5.2.1), i.e. "if there is a path connecting x and y." [1]

7.4. Let X be a simplicial set and x_0 a vertex of X. By definition, we call *Poincaré group* of X at x_0 the group $\Pi_1(X_1\, x_0) = \Pi_1(\Pi X_1\, x_0)$ (6.2). If $f\colon X \to Y$ is a morphism which sends x_0 to $f(x_0) = y_0$, we will write simply $\Pi_1(f, x_0)$ or $\Pi_1 f$ for the group homomorphism $\Pi_1(\Pi(f), x_0)$. Let then $.\Delta^\circ\mathcal{E}$ be the category of simplicial sets *with base point*, i.e. the category whose objects are the pairs $(X_1\, x_0)$ formed by a simplicial set X and one its vertices x_0; morphisms $f\colon (X_1\, x_0) \to (Y_1\, y_0)$ correspond to morphisms (also written $f\colon X \to Y$) of $\Delta^\circ\mathcal{E}$ such that $f(x_0) = y_0$.

It follows from the definition that Π_1 is a functor from $.\Delta^\circ\mathcal{E}$ to the category of groups.

[1] Un peu d'abstraction éloigne de la géométrie, beaucoup (?) d'abstraction y ramène.

Let $(X_1 x_0) \xleftarrow{f} (Z_1 z_0) \xrightarrow{g} (Y_1 y_0)$ be a diagram in $\Delta^\circ\mathscr{E}$. *Suppose that* X, Y, Z *are connected, and that the components* f_0 *and* g_0 *of* f *and* g *are injective.* Applying the functor Π_1 we obtain the following diagram in $\mathscr{G}r$:

$$\Pi(X) \xleftarrow{\Pi(f)} \Pi(Z) \xrightarrow{\Pi(g)} \Pi(Y).$$

According to 4.2 this diagram satisfies the hypothesis of lemma 6.2. Hence we have:

Proposition: ("Van Kampen" for simplicial sets.) *Under the above hypothesis, and if* t_0 *is the common image of both* x_0 *and* y_0 *in* $X \overset{Z}{\amalg} Y$, *we have:*

$$\Pi_1\left(X \overset{Z}{\amalg} Y, t_0\right) \approx \Pi_1(X, x_0) \overset{\Pi_1(Z, z_0)}{\amalg} \Pi_1(Y, y_0).$$

This follows from the fact that the square

$$\begin{array}{ccc} Z & \xrightarrow{f} & X \\ {\scriptstyle g}\downarrow & & \downarrow \\ Y & \rightarrow & X \overset{Z}{\amalg} Y \end{array}$$

is cocartesian; and Π transforms a cocartesian square into a cocartesian square.

7.5. We see easily that there are products in $\Delta^\circ(\mathscr{E})$. In fact, $(X \times Y, (x_0, y_0))$ is the product of $(X_1 x_0)$ and $(Y_1 y_0)$. In this paragraph we intend to show that Π_1 commutes with finite products. In order to do this, it is obviously sufficient to show that Π commutes with finite products.

Since the functor Π commutes with direct limits, it is sufficient to show that $\Pi(\Delta[n] \times \Delta[p])$ is identified with $\Pi\Delta[n] \times \Pi\Delta[p]$. For in that case, we have (see 1.1):

$$\begin{aligned} \Pi(X \times Y) &\approx \Pi\left(\varinjlim_{\alpha} d_X(\alpha) \times \varinjlim_{\beta} d_Y(\beta)\right) \approx \Pi \varinjlim_{\alpha,\beta} (d_X(\alpha) \times d_Y(\beta)) \\ &\approx \varinjlim_{\alpha,\beta} \Pi(d_X(\alpha) \times d_Y(\beta)) \approx \varinjlim_{\alpha,\beta} (\Pi\, d_X(\alpha) \times \Pi\, d_Y(\beta)) \\ &\approx \varinjlim_{\alpha} (\Pi\, d_X(\alpha)) \times \varinjlim_{\beta} (\Pi\, d_Y(\beta)) \approx \Pi X \times \Pi Y \end{aligned}$$

where $d_X \colon \Delta/X \to \Delta^\circ\mathscr{E}$ and $d_Y \colon \Delta/Y \to \Delta^\circ\mathscr{E}$ are the functors (1.1) which allow us to represent X and Y as direct limits of representable functors. We used the fact that in $\mathscr{C}at$, the product commutes with direct limits, which follows from the fact that the functor $\mathscr{C} \rightsquigarrow \mathscr{C} \times \mathscr{D}$ is left adjoint to the functor $\mathscr{F} \rightsquigarrow \mathscr{H}om(\mathscr{D}, \mathscr{F})$.

Now we check that the product of two simply connected groupoids is simply connected, and that the canonical morphism

$$\Pi(\Delta[n] \times \Delta[p]) \to \Pi\Delta[n] \times \Pi\Delta[p]$$

induces a bijection on objects and a surjection on morphisms. The groupoid on the right is simply connected; hence, in order to complete the demonstration, it is sufficient to show that $\Pi(\Delta[n]\times\Delta[p])$ is simply connected.

Let us go back to the description of $\Delta[n]\times\Delta[p]$ given in 5.5. The 1-simplices of $\Delta[n]\times\Delta[p]$ are in a one-to-one correspondence with the pairs (x, y) where $x\leqq y$ are points of $[n]\times[p]$. The 2-simplices of $\Delta[n]\times\Delta[p]$ are in a one-to-one correspondence with the triples (x, y, z) such that $x\leqq y\leqq z$ in $[n]\times[p]$. Hence a morphism $\alpha\colon a\to b$ of $\Pi\big(\Delta[n]\times\Delta(p)\big)$ may be represented by a path

$$(x_n, x_{n-i})\circ\cdots\circ(x_2, x_1)\circ(x_1, x_0)$$

where $x_0=a$, $x_n=b$ and $x_i\leqq x_{i+1}$ or $x_i\geqq x_{i+1}$. It is left to the reader to show, by induction, that this path is equivalent to the path

$$(b, y)\circ(y, a)$$

where y is the point (1.1). (Use the description of the relations given in 7.1, and the description of the 2-simplices of $\Delta[n]\times\Delta[p]$). Hence the morphism α is well defined by its domain and its range. Q.E.D.

7.6. In general, Π does not commute with kernels of pairs of morphisms as shown in the following example: consider the cocartesian square

$$\begin{array}{ccc}\dot{\Delta}[2] & \xrightarrow{i} & \Delta[2]\\ {\scriptstyle i}\downarrow & & \downarrow{\scriptstyle u}\\ \Delta[2] & \xrightarrow{v} & X\end{array}$$

where i is the canonical injection. The kernel of $\Delta[2]\overset{u}{\underset{v}{\rightrightarrows}}X$ is $\dot{\Delta}[2]$; but we have $\Pi\Delta[2]=\Pi X=Sc[2]$, and the kernel of $Sc[2]\rightrightarrows Sc[2]$ is $Sc[2]\neq\Pi\dot{\Delta}[2]$.

Chapter Three

Geometric Realization of Simplicial Sets

1. Geometric Realization of a Simplicial Set

1.1. First, let us give some general remarks which will divert us a little:

For each natural integer n, the set $\Delta([n],[1])$ will be totally ordered by saying that $f\geqq g$ if and only if $f(i)\leqq g(i)$ for each $i\in[n]$; moreover, we can identify the ordered set $\Delta([n],[1])$ with $[n+1]$ under the map $f\rightsquigarrow \operatorname{card} f^{-1}(0)$. We define thus a functor $[n]\rightsquigarrow\Delta([n],[1])$ from Δ to Δ°, which will be noted II, and which can be described as follows:

$$\mathrm{II}[n]=[n+1],\qquad \mathrm{II}\,\partial_n^i=\sigma_n^i,\qquad \mathrm{II}\,\sigma_n^i=\partial_{n+2}^{i+1}.$$

1.2. Let $\mathcal{Top}$ be the category of topological spaces, and $\boldsymbol{I}$ the segment $[0, 1]$ of $\boldsymbol{R}$. For each natural integer n, we will write $\boldsymbol{I}^{[n]}$ or $\boldsymbol{I}^{n+1}$ for the product of $n+1$ copies of the topological space $\boldsymbol{I}$. It is clear that this product depends on $[n]$ in a contravariant way (the set subordinated to the topological space $\boldsymbol{I}^{[n]}$ is identified with the set $\mathcal{E}([n], I)$ of maps of $[n]$ into $\boldsymbol{I}$), so that we have a simplicial topological space $[n] \rightsquigarrow \boldsymbol{I}^{[n]}$ which will be noted $\boldsymbol{I}^?$. The composition

$$\Delta \xrightarrow{\text{II}} \Delta^\circ \xrightarrow{\boldsymbol{I}^?} \mathcal{Top}$$

is then a cosimplicial topological space, which we will write III, and which can be described as follows:

$\text{III}^n = \boldsymbol{I}^{n+2}$

${}^{\text{III}}\partial_n^i : \boldsymbol{I}^{n+1} \to \boldsymbol{I}^{n+2}$ is the map

$$(t_0, t_1, \ldots, t_n) \rightsquigarrow (t_0, \ldots, t_{i-1}, t_i, t_i, t_{i+1}, \ldots, t_n)$$

${}^{\text{III}}\sigma_n^i : \boldsymbol{I}^{n+3} \to \boldsymbol{I}^{n+2}$ is the map

$$(t_0, t_1, \ldots, t_n, t_{n+1}) \rightsquigarrow (t_0, \ldots, t_i, t_{i+2}, \ldots, t_{n+2}).$$

From now on, we will write Δ^n *for the subspace of* $\boldsymbol{I}^{n+2}$ *formed by all points* $(t_0, t_1, \ldots, t_n, t_{n+1})$, *such that* $0 = t_0 \leqq t_1 \leqq \cdots \leqq t_n \leqq t_{n+1} = 1$. The sequence $\Delta^0, \Delta^1 \ldots$ is obviously a sub-functor of III, which is written $\Delta^?: \Delta \to \mathcal{Top}$.

1.3. The first part of our work will be to determine the group of automorphisms of the functor $\Delta^?$: let $\mathfrak{G}$ be the group of homeomorphisms s of the segment $[0, 1]$ such that $s(0) = 0$ [and hence $s(1) = 1$]. The group $\mathfrak{G}$ obviously operates on the functor III; more precisely, each $s \in \mathfrak{G}$ induces an automorphism s^n: $(t_0, t_1, \ldots, t_n, t_{n+1}) \rightsquigarrow (st_0, st_1, \ldots, st_n, st_{n+1})$ of $\text{III}^n = \boldsymbol{I}^{n+2}$ which is compatible with the face and degeneracy operators. Moreover, such an s is an increasing map, and hence, it fixes subset Δ^n of $\boldsymbol{I}^{n+2}$ formed by all points $(t_0, t_1, \ldots, t_n, t_{n+1})$ such that $0 = t_0 \leqq t_1 \leqq \cdots \leqq t_n \leqq t_{n+1}$. Thus we obtain a homomorphism $s \rightsquigarrow s^?$ from $\mathfrak{G}$ into the group of automorphisms of $\Delta^?$.

Proposition: *The monomorphism* $s \rightsquigarrow s^?$ *is an isomorphism from the group of homeomorphisms* s *of* $\boldsymbol{I}$ *such that* $s(0) = 0$ *onto the group of automorphisms of the functor* $\Delta^?$.

It is clear that the homomorphism $s \rightsquigarrow s^?$ is injective; let us show that it is surjective: let $s_0, s_1, \ldots$ be continuous automorphisms of $\Delta^0, \Delta^1, \ldots$ compatible with face and degeneracy operators. Since s_1 is compatible with ∂_1^0: $\Delta^0 \to \Delta^1$, we have $s_1(0, 0, 1) = (0, 0, 1)$, so that there exists an $s \in \mathfrak{G}$ such that $s_1 = s^1$, with the above notations. On the other hand, when σ suns through the epimorphisms $[n] \to [1]$ of Δ, Δ^σ: $\Delta^n \to \Delta^1$ runs through the maps $(0, t_1, \ldots, t_n, 1) \rightsquigarrow (0, t_i, 1)$, $1 \leqq i \leqq n$, and we have $\Delta^\sigma \circ s_n = s_1 \circ \Delta^\sigma$. Hence $s_n(0, t_1, \ldots, t_n, 1) = (0, st_1, \ldots, st_n, 1)$, i.e. $s_n = s^n$. Q.E.D.

1.4. Now let us apply proposition II, 1.3 to the present case: the functor $\Delta^?: \Delta \to \mathcal{T}op$ defines a pair of adjoint functors $S: \mathcal{T}op \to \Delta^\circ \mathcal{E}$ and $|?|: \Delta^\circ \mathcal{E} \to \mathcal{T}op$.

The first one is the *singular complex functor*, which associates with each topological space T the complex $[n] \rightsquigarrow (S\,T)_n = \mathcal{T}op(\Delta^n, T)$, also called the *complex of singular simplices of* T. The other one is left adjoint to the first; the image $|X|$ of a complex X under $|?|$ is the direct limit of the functor

$$\delta_X: \Delta[n] \xrightarrow{\alpha} X \rightsquigarrow \Delta^n$$

from Δ/X into $\mathcal{T}op$ (see II, 1.3). Usually, we say that $|X|$ is the *geometric realization* of the complex X. Recall that the "geometric realization" functor is characterized (up to a functor isomorphism) by the following two conditions: $|?|$ commutes with direct limits; moreover, if $h^\Delta: \Delta \to \Delta^\circ \mathcal{E}$ is the functor $[n] \rightsquigarrow \Delta[n]$, the composition $|?| \circ h^\Delta$ is identified with $\Delta^?: \Delta \to \mathcal{T}op$. In particular, the realization $|\Delta[n]|$ of the standard n-simplex is identified with the geometric simplex of dimension n, i.e. Δ^n.

1.5. In a similar way as in II, 4.2, let us now determine the functor $|?|$. Again, take the diagram of II, 3.9

$$(*) \qquad \coprod_{0 \le i < j \le n} \Delta[n-2]_{i,j} \underset{v}{\overset{u}{\rightrightarrows}} \coprod_{0 \le i \le n} \Delta[n-1]_i \xrightarrow{p} \Delta[n].$$

Consider now the geometric realization of the sequence $(*)$: the restriction of $|p|$ to $|\Delta[n-1]_i|$ is a homeomorphism f_i from $|\Delta[n-1]_i| \simeq \Delta^{n-1}$ onto the i-th face Δ^n_i of Δ^n [Δ^n_i is the intersection of $\Delta^n \subset R^{n+2}$ with the set of points $(t_0, \ldots, t_{n+1})$ such that $t_i = t_{i+1}$]. If we compose f_i with $|\Delta(\partial^j_{n-1})|: |\Delta[n-2]_{i,j}| \to |\Delta[n-1]_i|$, we get a homeomorphism $f_{i,j}$ from $|\Delta[n-2]_{i,j}|$ onto $\Delta^n_i \cap \Delta^n_j$. We then have the following commutative diagram (Fig. 12).

$$\begin{array}{ccccc}
\coprod |\Delta[n-2]_{i,j}| & \underset{|v|}{\overset{|u|}{\rightrightarrows}} & \coprod |\Delta[n-1]_i| & \xrightarrow{|p|} & |\Delta[n]| \\
\downarrow{\scriptstyle \coprod f_{i,j}} & & \downarrow{\scriptstyle \coprod f_i} & & \downarrow{\scriptstyle \mathrm{Id}} \\
\coprod \Delta^n_i \cap \Delta^n_j & \underset{v'}{\overset{u'}{\rightrightarrows}} & \coprod \Delta^n_i & \xrightarrow{p'} & \Delta^n \simeq |\Delta[n]|
\end{array}$$

Fig. 12

where p' is defined by the inclusions of Δ^n_i into Δ^n, and where u' (resp. v') is defined by the inclusions of $\Delta^n_i \cap \Delta^n_j$ into Δ^n_i (resp. into Δ^n_j).

It is clear that p' induces a homeomorphism from coker (u', v') onto the union of the faces of Δ^n (the boundary of Δ^n). Since $|?|$ transforms cokernels into cokernels, and since $\coprod f_i$ and $\coprod f_{i,j}$ are homeomorphisms, we have the following proposition:

Proposition: If i is the inclusion of $\dot{\Delta}[n]$ into $\Delta[n]$, $|i|$ is a homeomorphism from the geometric realization $|\dot{\Delta}[n]|$ of the boundary of $\Delta[n]$ onto the boundary $\dot{\Delta}^n$ of the geometric realization Δ^n of $\Delta[n]$.

Let X be a simplicial set. We know that X is equal to the direct limit $\varinjlim_n Sk^n X$, and hence, that $|X|$ is equal to the direct limit $\varinjlim |Sk^n X|$, which we will now describe:

$Sk^0 X$ is identified with the direct sum of a family of standard simplice $\Delta[0]$ indexed by the vertices of X; since the geometric realization of $\Delta[0]$ is the point Δ^0, we see that $|Sk^0 X|$ is identified with the topological direct sum of points indexed by the vertices of X; in other words, $|Sk^0 X|$ is identified with X_0, with the discrete topology.

Suppose that $|Sk^{n-1} X|$ is given, and let us determine $|Sk^n X|$; since the geometric realization functor commutes with amalgamated sums and direct sums, it follows from II, 3.8 that $|Sk^n X|$ is identified with the amalgamated sum of the diagram of Fig. 13

$$\begin{array}{ccc} \coprod_{\sigma \in \Sigma^n} |\dot{\Delta}[n]_\sigma| & \xrightarrow{p} & |Sk^{n-1} X| \\ {\scriptstyle |\text{inclusion}|}\downarrow & & \\ \coprod_{\sigma \in \Sigma^n} |\Delta[n]_\sigma| & & \end{array}$$

Fig. 13

the restriction of p to $|\dot{\Delta}[n]_\sigma|$ being defined by the singular simplex $\bar{\bar{\sigma}}: \Delta[n] \to X$. Moreover, $|\Delta[n]_\sigma|$ is a geometric simplex of dimension n, say Δ^n_σ, and $|\dot{\Delta}[n]_\sigma|$ is identified with the boundary $\dot{\Delta}^n_\sigma$ of Δ^n_σ. Hence we have a cocartesian square (Fig. 14)

$$\begin{array}{ccc} \coprod_{\sigma \in \Sigma^n} \dot{\Delta}^n_\sigma & \longrightarrow & |Sk^{n-1} X| \\ {\scriptstyle \text{inclusion}}\downarrow & & \downarrow \\ \coprod_{\sigma \in \Sigma^n} \Delta^n_\sigma & \longrightarrow & |Sk^n X| \end{array}$$

Fig. 14

which means that $|Sk^n X|$ is obtained by attaching to the topological space $|Sk^{n-1} X|$ a family (Δ^n_σ) of geometric n-simplices along their boundaries. It is equivalent to say that the map $|Sk^{n-1} X| \to |Sk^n X|$ is a homeomorphism from $|Sk^{n-1} X|$ onto a closed subspace of $|Sk^n X|$ (from now on, $|Sk^{n-1} X|$ will be identified with this subspace); that the complement of $|Sk^{n-1} X|$ in $|Sk^n X|$ is identified with the topological direct sum of the interiors $\mathring{\Delta}^n_\sigma$ of the geometric simplices Δ^n_σ; that a subspace U of $|Sk^n X|$ is open if $U \cap |Sk^{n-1} X|$ is open in $|Sk^{n-1} X|$ and if the inverse image of U in each Δ^n_σ is open. In particular, the continuous map $|\tilde{\sigma}|: \Delta^n_\sigma \to |Sk^n X|$ induces a homeomorphism from $\mathring{\Delta}^n_\sigma$ onto a connected component of $|Sk^n X| - |Sk^{n-1} X|$; we will call this connected component *the cell of dimension n and index σ.*

1.6. It is a well known fact that two functors adjoint to each other, in particular S and $|?|$, have "the same" group of automorphisms. According to II, 1.3, the group of automorphisms of $|?|$ is identified with the group of automorphisms of $\Delta^?$, i.e. with the group $\mathfrak{G}$ of strictly increasing continuous maps $s: I \to I$ such that $s(0)=0$ and $s(1)=1$. We will see how this group operates on the geometric realization $|X|$ of a complex X.

The operations of $\mathfrak{G}$ in $|Sk^n X|$ and $|X|$ are obviously compatible with the inclusion of $|Sk^n X|$ into $|X|$. Hence $\mathfrak{G}$ preserves each connected component of $|Sk^n X| - |Sk^{n-1} X|$, i.e. each cell of dimension n and index σ. Since the continuous map $|\tilde{\sigma}|: \Delta^n_\sigma \to |X|$ is compatible with the operations of $\mathfrak{G}$ in Δ^n_σ and $|X|$, and since $\mathfrak{G}$ operates transitively in $\mathring{\Delta}^n_\sigma$, as easily seen by the formula $s(t_0, \ldots, t_{n+1}) = (st_0, \ldots, st_{n+1})$, it follows that $\mathfrak{G}$ operates transitively in each cell of $|X|$:

Proposition: For each complex X, the cells of |X| are the orbits of the group of automorphisms of the geometric realization functor.

In particular, if $f: X \to Y$ is a morphism of $\Delta^\circ\mathscr{E}$, $|f|: |X| \to Y$ is compatible with the operations of $\mathfrak{G}$ on $|X|$ and $|Y|$. Hence $|f|$ transforms the cells of $|X|$ into cells of $|Y|$.

1.7. We will now construct a basis of open sets for $|X|$. In order to do this, let us show that each open set $\boldsymbol{U}_{n-1}$ of $|Sk^{n-1}X|$ can be extended to $|Sk^n X|$: let $\lambda = (\lambda_\sigma)$ be a family of numbers belonging to the half open interval $]0, 1]$, and indexed by the non degenerate simplices σ of X; for each $\sigma \in X_n$, let G_σ be the barycenter of the geometric simplex Δ^n_σ, and $\boldsymbol{U}_\sigma$ the set of points of Δ^n_σ of the form $tG_\sigma + (1-t)M$, $0 \leqq t < \lambda_\sigma$, where M runs through the inverse image $|\tilde{\sigma}|^{-1}(\boldsymbol{U}_{n-1})$ of $\boldsymbol{U}_{n-1}$ in $\mathring{\Delta}^n_\sigma$ (notations of 1.5); if $\boldsymbol{U}_n$ denotes the union of $\boldsymbol{U}_{n-1}$ with the images $|\tilde{\sigma}|(\boldsymbol{U}_\sigma)$, for all $\sigma \in \Sigma^n$, it is clear that $\boldsymbol{U}_n$ intersects $|Sk^{n-1}X|$ along $\boldsymbol{U}_{n-1}$, and that we have $|\tilde{\sigma}|^{-1}(\boldsymbol{U}_n) = \boldsymbol{U}_\sigma$ for each $\sigma \in \Sigma^n$; according to 1.5, $\boldsymbol{U}_n$ is then open in $|Sk^n X|$; moreover, let $\boldsymbol{V}$ be a subset of $\boldsymbol{U}_n$, such that the intersection $\boldsymbol{V} \cap \boldsymbol{U}_{n-1}$ (resp. the inverse image $\boldsymbol{V}_\sigma$ of $\boldsymbol{V}$ in $\boldsymbol{U}_\sigma$) is open in $\boldsymbol{U}_{n-1}$ (resp. in $\boldsymbol{U}_\sigma$, $\forall \sigma \in \Sigma^n$); then $\boldsymbol{V} \cap |Sk^{n-1}X|$ (resp. $|\tilde{\sigma}|^{-1}(\boldsymbol{V})$) coincides with $\boldsymbol{V} \cap \boldsymbol{U}_{n-1}$ (resp. with $\boldsymbol{V}_\sigma$); consequently, $\boldsymbol{V}$ is open in $|Sk^n X|$, and hence in $\boldsymbol{U}_n$. This means that we have a cocartesian square (Fig. 15)

$$\begin{array}{ccc} \coprod_{\sigma\in\Sigma^n} |\tilde{\sigma}|^{-1}(\boldsymbol{U}_{n-1}) & \longrightarrow & \boldsymbol{U}_{n-1} \\ \downarrow & & \downarrow \\ \coprod_{\sigma\in\Sigma^n} \boldsymbol{U}_\sigma & \longrightarrow & \boldsymbol{U}_n \end{array}$$

Fig. 15

where vertical arrows are inclusions, the horizontal ones being induced by the singular simplices $\tilde{\sigma}$.

We can now do this construction another time, replacing n by $n+1$ and U_{n-1} by U_n. We construct thus a sequence $U_{n-1} \subset U_n \subset U_{n+1} \subset \cdots$ of subsets of $|X|$ which have the following property: U_r is open in $|Sk^r X|$, and we have $U_r = |Sk^r X| \cap U_{r+1}$. Since $|X|$ is the topological direct limit of the $|Sk^r X|$, the union U_{n-1}^λ of the U_r is an open set of $|X|$ which will be called *the extension of width λ of U_{n-1} to $|X|$*. This extension is compatible with intersection: if U_{n-1} and V_{n-1} are two open sets of $|Sk^{n-1}X|$, we have $(U_{n-1} \cap V_{n-1})^\lambda = U_{n-1}^\lambda \cap V_{n-1}^\lambda$. In particular, if U_{n-1} and V_{n-1} are disjoint, U_{n-1}^λ and V_{n-1}^λ are also disjoint.

1.8. Still using the notation of 1.7, let x be a point of $|X|$ belonging to a cell of dimension n and index σ; let ε be a number belonging to $]0, 1[$, and $\mathcal{O}_x$ the point of Δ_σ^n whose image under the map $|\tilde{\sigma}| : \Delta_\sigma^n \to |X|$ is x; finally, let V_x^ε be the image in $|X|$ of the set of points of Δ_σ^n of the form $(1-t)\,\mathcal{O}_x + tM$, $0 \leqq t < \varepsilon$, where M runs through the boundary $\dot{\Delta}_\sigma^n$ of Δ_σ^n. This image V_x^ε is obviously an open neighborhood of x in $|Sk^n X|$. Moreover, it is clear that the extension $V_x^{\varepsilon,\lambda}$ of width λ of V_x^ε to $|X|$ runs through a basis of neighborhoods of x in $|X|$ when ε and λ vary.

Proposition: The geometric realization of a simplicial set is a Hausdorff space.

Let x and y be distinct points of $|X|$ belonging to cells of dimensions m and p respectively. If $m < p$, the extension of width λ of V_x^ε to $|Sk^p X|$ is disjoint from V_y^ε for a well chosen ε. Hence, according to 1.7, we can choose ε so that $V_x^{\varepsilon,\lambda}$ and $V_y^{\varepsilon,\lambda}$ are disjoint. Similarly, if $m = p$, x and y belong to $|Sk^m X| - |Sk^{m-1} X|$, which is an open subset of $|Sk^m X|$ isomorphic to the topological direct sum of the spaces $\mathring{\Delta}_\sigma^m$, $\sigma \in \Sigma^m$. For a well chosen ε, V_x^ε and V_y^ε do not intersect, so that $V_x^{\varepsilon,\lambda}$ and $V_y^{\varepsilon,\lambda}$ are again disjoint.

1.9. Now let us use again the notations of 1.7. If U_{n-1} is open in $|Sk^{n-1}X|$, we will see that U_{n-1} is *a deformation retract of the open subset U_n of $|Sk^n X|$ constructed in* 1.7: let r_σ: $I \times U_\sigma$ be the retracting deformation of U_σ onto $|\tilde{\sigma}|^{-1}(U_{n-1})$ such that $r_\sigma(1-s, tG_\sigma + (1-t)M) = st\,G_\sigma + (1-st)M$ (I is the interval $[0, 1]$). Since $I \times U_n$ is identified with the amalgamated sum of $I \times U_{n-1}$ and $\coprod_\sigma I \times U_\sigma$ under $\coprod_\sigma I \times |\tilde{\sigma}|^{-1}(U_{n-1})$ (see 1.7 and 2.1 below), there is a continuous map r: $I \times U_n \to U_n$, which induces on $I \times U_{n-1}$ the canonical projection onto U_{n-1}, and on $I \times U_\sigma$, the composition $(s, x) \rightsquigarrow |\tilde{\sigma}|\,(r_\sigma(s, x))$. This map r is a retracting deformation of U_n onto U_{n-1}.

It follows that U_{n-1} is a *deformation retract of the extension U_{n-1}^λ of U_{n-1} to $|X|$ of width λ*. For since U_{n-1}^λ is the direct limit of the spaces

U_r, $r \geqq n-1$, we can define a retracting deformation $(s, x) \rightsquigarrow \varrho(s, x)$ of U_{n-1}^λ onto U_{n-1}, by retracting U_n onto U_{n-1} while s runs through the interval $[\frac{1}{2}, 1]$, by retracting U_{n+1} onto U_n while s runs through the interval $[\frac{1}{4}, \frac{1}{2}]$, by retracting U_{n+2} onto U_{n+1} while $s \in [\frac{1}{8}, \frac{1}{4}]$ ....

If we apply this construction to the case where U_{n-1} is one of the open sets $V_{x,i}^{\varepsilon,\lambda}$ of 1.8, we see that in particular *each point x of $|X|$ has a neighborhood-basis consisting of contractible open sets.*

2. Kelley Spaces

2.1. If Y and Z are topological spaces, we write $CO(Y, Z)$ for the set of continuous maps from Y to Z, with the compact-open topology (see for example HILTON-WYLIE, Homology theory, p. 286).

Suppose that Y is *locally compact.* If, for each map $f: X \times Y \to Z$ and for each $x \in X$, f_x denotes the function $y \rightsquigarrow f(x, y)$, we know that the map $f \rightsquigarrow (f_x)_{x \in X}$ is a bijection

$$\mathcal{T}op(X \times Y, Z) \simeq \mathcal{T}op(X, CO(Y, Z))$$

and hence that the functor $X \rightsquigarrow X \times Y$ is left adjoint to the functor $Z \rightsquigarrow CO(Y, Z)$ and commutes with direct limits.

Now if we suppose that X is a Kelley space (I, 1.5.3), X is the topological direct limit of its compact subsets X'; hence $X \times Y$ is the direct limit in $\mathcal{T}op$ of the Kelley spaces $X' \times Y$; since $X \times Y$ is Hausdorff, it is a Kelley space, and we have:

2.1.1. *If X is a Kelley space and Y is locally compact, the equality $X \times Y = (X \times Y)_{\mathcal{K}e}$ holds* (see I, 1.5.3).

2.1.2. If Y is a Kelley space, $CO(Y, Z)$ is identified with the topological inverse limit in $\mathcal{T}op$ of the spaces $CO(Y', Z)$, where Y' runs through the compact subsets of Y. Suppose then that X, Y, Z *are Kelley spaces.* Then $(X \times Y)_{\mathcal{K}e}$ is the topological direct limit of the spaces $X' \times Y'$ where X' and Y' run through the compact subsets of X and Y respectively; we then have the following identifications:

$$\begin{aligned}
\mathcal{K}e((X \times Y)_{\mathcal{K}e}, Z) &\simeq \mathcal{T}op\left(\varinjlim_{X', Y'} X' \times Y', Z\right) \simeq \varprojlim_{X', Y'} \mathcal{T}op(X' \times Y', Z) \\
&\simeq \varprojlim_{X', Y'} \mathcal{T}op(X', CO(Y', Z)) \simeq \varprojlim_{X'} \mathcal{T}op\left(X', \varprojlim_{Y'} CO(Y', Z)\right) \\
&\simeq \mathcal{T}op\left(\varinjlim_{X'} X', CO(Y, Z)\right) \simeq \mathcal{T}op(X, CO(Y, Z)) \\
&\simeq \mathcal{K}e(X, CO_{\mathcal{K}e}(Y, Z))
\end{aligned}$$

where $CO_{\mathcal{K}e}(Y, Z)$ denotes the Kelley space $(CO(Y, Z))_{\mathcal{K}e}$ associated with the Hausdorff space $CO(Y, Z)$. Thus we have proved the following proposition:

Proposition: In the category of Kelley spaces, the functor $\boldsymbol{X} \rightsquigarrow (\boldsymbol{X}\times\boldsymbol{Y})_{\mathcal{K}e}$ *is left adjoint to the functor* $\boldsymbol{Z} \rightsquigarrow \boldsymbol{CO}_{\mathcal{K}e}(\boldsymbol{Y},\boldsymbol{Z})$. *Hence the functor* $\boldsymbol{X} \rightsquigarrow (\boldsymbol{X}\times\boldsymbol{Y})_{\mathcal{K}e}$ *commutes with direct limits.*

2.2. If $\boldsymbol{B}$ is a Kelley space, $\mathcal{K}e/\boldsymbol{B}$ denotes, as usual, the category whose objects are the morphisms $\xi\colon \boldsymbol{X}\to\boldsymbol{B}$ of range $\boldsymbol{B}$ of $\mathcal{K}e$; if $\xi\colon \boldsymbol{X}\to\boldsymbol{B}$ and $\eta\colon \boldsymbol{Y}\to\boldsymbol{B}$ are two such objects, a morphism from the first to the second is a morphism $\alpha\colon \boldsymbol{X}\to\boldsymbol{Y}$ of $\mathcal{K}e$ such that $\xi=\eta\circ\alpha$. Sometimes, we say that $\boldsymbol{X}$ is an object over $\boldsymbol{B}$, instead of saying that ξ is an object of $\mathcal{K}e/\boldsymbol{B}$; we also write $\boldsymbol{X}$ instead of ξ, and we say that ξ is the structural morphism from $\boldsymbol{X}$ to $\boldsymbol{B}$.

Now let us consider two objects over $\boldsymbol{B}$: $\boldsymbol{f}\colon \boldsymbol{B}'\to\boldsymbol{B}$ and $\xi\colon \boldsymbol{X}\to\boldsymbol{B}$. The canonical projection of the fibred product $(\boldsymbol{B}'\times_{\boldsymbol{B}}\boldsymbol{X})_{\mathcal{K}e}$ onto the first factor makes $(\boldsymbol{B}'\times_{\boldsymbol{B}}\boldsymbol{X})_{\mathcal{K}e}$ an object over $\boldsymbol{B}'$; we say that the functor from $\mathcal{K}e/\boldsymbol{B}$ to $\mathcal{K}e/\boldsymbol{B}'$ which associates with each object $\boldsymbol{X}$ over $\boldsymbol{B}$ the object $(\boldsymbol{X}\times_{\boldsymbol{B}}\boldsymbol{B}')_{\mathcal{K}e}$ is the *change of base functor* defined by $\boldsymbol{f}$. For example, if $\boldsymbol{B}$ reduces to a point, $(\boldsymbol{B}'\times_{\boldsymbol{B}}\boldsymbol{X})_{\mathcal{K}e}$ is identified with $(\boldsymbol{B}'\times\boldsymbol{X})_{\mathcal{K}e}$; in this case, the previous proposition shows that the change of base functor commutes with direct limits. "For want of a better one", we will use in general the following proposition:

Proposition: Let $\boldsymbol{B}$ *be a Kelley space,* $\boldsymbol{f}\colon \boldsymbol{B}'\to\boldsymbol{B}$ *an object of* $\mathcal{K}e/\boldsymbol{B}$, i *the injection of* $\mathcal{K}e$ *into* $\mathcal{T}op$ *and* $\delta\colon T\to\mathcal{K}e/\boldsymbol{B}$ *a diagram which associates with* $t\in\mathfrak{Ob}\,T$ *the object* $\delta(t)\colon \boldsymbol{d}(t)\to\boldsymbol{B}$ *of* $\mathcal{K}e/\boldsymbol{B}$. *Suppose that* $\varinjlim i\circ\boldsymbol{d}$ *is a Hausdorff space. Then the canonical morphism*

$$\varinjlim_{t\in T}\left(\boldsymbol{d}(t)\times_{\boldsymbol{B}}\boldsymbol{B}'\right)_{\mathcal{K}e}\to\left(\Big(\varinjlim_{t\in T}\boldsymbol{d}(t)\Big)\times_{\boldsymbol{B}}\boldsymbol{B}'\right)_{\mathcal{K}e}$$

of $\mathcal{K}e$ *is a homeomorphism.*

We know indeed that $(\boldsymbol{d}(t)\times_{\boldsymbol{B}}\boldsymbol{B}')_{\mathcal{K}e}$ is the kernel of the pair of morphisms $\delta(t)\circ\mathrm{pr}_1, \boldsymbol{f}\circ\mathrm{pr}_2\colon (\boldsymbol{d}(t)\times\boldsymbol{B}')_{\mathcal{K}e}\rightrightarrows\boldsymbol{B}$, where pr_1 and pr_2 are the canonical projections of the product onto its factors. Passing to the direct limit in $\mathcal{K}e$, we get the following diagram (Fig. 16):

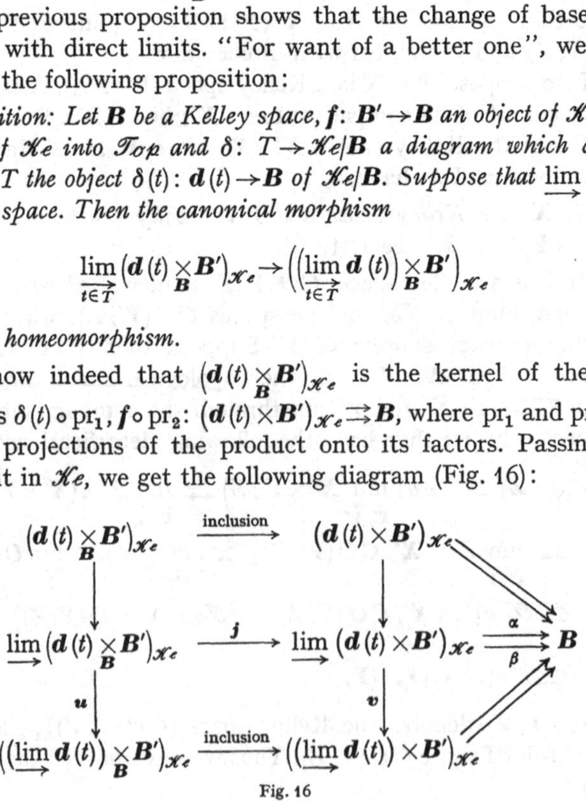

Fig. 16

where all morphisms are the "obvious" ones. By 2.1, $\boldsymbol{v}$ is invertible, and we want to show that $\boldsymbol{u}$ is also invertible; since $((\varinjlim \boldsymbol{d}(t)) \times_{\boldsymbol{B}} \boldsymbol{B}')_{\mathcal{K}e}$ is the kernel of the pair of morphisms $((\varinjlim \boldsymbol{d}(t)) \times \boldsymbol{B}')_{\mathcal{K}e} \rightrightarrows \boldsymbol{B}$, it is sufficient to show that $\boldsymbol{j}$ is a closed injection whose point set image is $\operatorname{Ker}(\alpha, \beta)$.

Since $\boldsymbol{v}$ is invertible and since the direct limit of the sets subordinated to $\boldsymbol{d}(t)$ is the set subordinated to $\varinjlim \boldsymbol{d}(t)$, we see that the set subordinated to $\varinjlim (\boldsymbol{d}(t) \times \boldsymbol{B}')_{\mathcal{K}e}$ is the direct limit of the sets subordinated to $(\boldsymbol{d}(t) \times \boldsymbol{B}')_{\mathcal{K}e}$. Let L be the direct limit of the sets subordinated to $(\boldsymbol{d}(t) \times_{\boldsymbol{B}} \boldsymbol{B}')_{\mathcal{K}e}$. Then, passing to the limit, the inclusions $(\boldsymbol{d}(t) \times_{\boldsymbol{B}} \boldsymbol{B}')_{\mathcal{K}e} \to (\boldsymbol{d}(t) \times \boldsymbol{B}')_{\mathcal{K}e}$ define an injection of L into $\varinjlim (\boldsymbol{d}(t) \times \boldsymbol{B}')_{\mathcal{K}e}$; this injection is the composition of the canonical surjection $\boldsymbol{p}: L \to \varinjlim (\boldsymbol{d}(t) \times_{\boldsymbol{B}} \boldsymbol{B}')_{\mathcal{K}e}$ with $\boldsymbol{j}$. Hence $\boldsymbol{p}$ is a bijection, $\boldsymbol{j}$ is an injection, and $\boldsymbol{u}$ is a bijection (direct limits commute with changes of base in the category of sets.)

Now let $\boldsymbol{F}$ be a closed subset of $\varinjlim (\boldsymbol{d}(t) \times_{\boldsymbol{B}} \boldsymbol{B}')_{\mathcal{K}e}$. The "point-set" equalities $(\boldsymbol{d}(t) \times_{\boldsymbol{B}} \boldsymbol{B}')_{\mathcal{K}e} = \operatorname{Ker}(\delta(t) \circ \mathrm{pr}_1, \boldsymbol{f} \circ \mathrm{pr}_2)$ and $\operatorname{Im} \boldsymbol{j} = \operatorname{Ker}(\alpha, \beta)$ imply that the inverse image $\boldsymbol{F}'$ of $\boldsymbol{j}(\boldsymbol{F})$ in $(\boldsymbol{d}(t) \times \boldsymbol{B}')_{\mathcal{K}e}$ coincides with the inverse image $\boldsymbol{F}''$ of $\boldsymbol{F}$ in $(\boldsymbol{d}(t) \times_{\boldsymbol{B}} \boldsymbol{B}')_{\mathcal{K}e}$; since $\boldsymbol{F}$ is closed, $\boldsymbol{F}''$, and hence $\boldsymbol{F}'$, are closed for all t. This means that $\boldsymbol{j}(\boldsymbol{F})$ is closed.

3. Exactness Properties of the Geometric Realization Functor

3.1. Let us go back now to the geometric realization functor: if X is a complex, $|X|$ is the direct limit of the functor

$$\delta_X : \Delta[n] \xrightarrow{\alpha} X \rightsquigarrow \Delta^n$$

from Δ/X to $\mathcal{T}op$ (see II, 1.3). Hence $|X|$ is Hausdorff (§ 1.8) and is a direct limit of Kelley spaces, i.e. a Hausdorff quotient of a direct sum of Kelley spaces; by (I, 1.5.3), *$|X|$ is a Kelley space.* We will use this fact to change our definitions slightly: except when it is charly specified that we use the first definition, we will suppose from now on that *the domain of the geometric realization functor is $\Delta^\circ\mathcal{E}$ and its range is the category $\mathcal{K}e$ of Kelley spaces.* With this convention, we have the following theorem:

Theorem: The geometric realization functor $|?|: \Delta^\circ\mathcal{E} \to \mathcal{K}e$ commutes with finite inverse limits and with directs limits; moreover, it is conservative (i.e. *a morphism α of $\Delta^\circ\mathcal{E}$ is invertible if $|\alpha|$ is invertible*).

It follows in particular from this theorem that the functor $|?|$: $\Delta^\circ\mathcal{E} \to \mathcal{K}e$ transforms a product into a product, a cartesian square (resp. cocartesian) into a cartesian square (resp. cocartesian), and that it commutes with the kernel (resp. the cokernel) of a pair of morphisms.

Consider for instance a morphism $f\colon X \to Y$ of $\Delta^\circ\mathscr{E}$ and the cartesian and cocartesian squares of Fig. 17:

$$
\begin{array}{ccc}
X \prod_Y X & \xrightarrow{\mathrm{pr}_2} & X \\
\downarrow{\scriptstyle \mathrm{pr}_1} & & \downarrow{\scriptstyle f} \\
X & \xrightarrow{f} & Y
\end{array}
\qquad\qquad
\begin{array}{ccc}
X & \xrightarrow{f} & Y \\
\downarrow{\scriptstyle f} & & \downarrow{\scriptstyle \mathrm{in}_2} \\
Y & \xrightarrow{\mathrm{in}_1} & Y \coprod_X Y
\end{array}
$$

Fig. 17

It is clear that f is the composition of the canonical projection of X onto coker $(\mathrm{pr}_1, \mathrm{pr}_2)$, with an isomorphism from coker $(\mathrm{pr}_1, \mathrm{pr}_2)$ onto $\ker(\mathrm{in}_1, \mathrm{in}_2)$ and with the canonical monomorphism from $\ker(\mathrm{in}_1, \mathrm{in}_2)$ into Y. Hence the continuous map $|f|\colon |X| \to |Y|$ is the composition of the canonical projection of $|X|$ onto coker $(|\mathrm{pr}_1|, |\mathrm{pr}_2|)$ (which is identified with $|\mathrm{coker}\ (\mathrm{pr}_1, \mathrm{pr}_2)|$) with an isomorphism from coker $(|\mathrm{pr}_1|, |\mathrm{pr}_2|)$ onto $\ker(|\mathrm{in}_1|, |\mathrm{in}_2|)$ (which is a closed subspace of $|Y|$ that can be identified with $|\ker(\mathrm{in}_1, \mathrm{in}_2)|$), and with the inclusion of $\ker(|\mathrm{pr}_1|, |\mathrm{pr}_2|)$ into $|Y|$.

The above remarks show then that the functor $|?|$ also commutes with the image formation of a morphism. We see easily, for instance, that the boundary $\dot{\Delta}[n]$ of $\Delta[n]$ is the image of the morphism p of 1.5; similarly, the boundary $\dot{\Delta}^n$ of the geometric simplex Δ^n is the image of $|p|$. We then find again the isomorphism $|\dot{\Delta}[n]| \simeq \dot{\Delta}^n$ of 1.5.

These few remarks show the importance of the theorem given above. The demonstration of this theorem will fill the rest of paragraph 3; but first, we have unfortunately to prove some of the corollaries given above. Let us show first that $|?|$ commutes with direct limits:

Let i be the inclusion of $\mathscr{K}e$ into $\mathscr{T}op$. We know already thet $i \circ |?|$ commutes with direct limits; in other words, for each diagram $d\colon T \to \Delta^\circ\mathscr{E}$, $\varinjlim i \circ |?| \circ d$ is identified with $i\,|\varinjlim d|$; since this space is HAUSDORFF, $\varinjlim i \circ |?| \circ d$ is identified with $i\,(\varinjlim |?| \circ d)$, by I, 1.5.3. Consequently, $i\,(\varinjlim |?| \circ d)$ is identified with $i\,|\varinjlim d|$, and $\varinjlim |?| \circ d$ with $|\varinjlim d|$.

3.2. Let $f\colon X \to Y$ be a monomorphism between complexes. *We will show that* $|f|\colon |X| \to |Y|$ *is a closed injection.* Of course, we may restrict ourselves to the case where f is the inclusion of a subcomplex X of Y into Y. Let then Σ'^n be the set of non degenerate n-simplices of Y which do not belong to X. The argument used in II, 3.8 shows that we have a cocartesian square (Fig. 18),

$$
\begin{array}{ccc}
\coprod_{\sigma \in \Sigma'^n} \dot{\Delta}[n]_\sigma & \longrightarrow & Sk^{n-1} Y \cup X \\
\downarrow{\scriptstyle i} & & \downarrow{\scriptstyle f_{n-1}} \\
\coprod_{\sigma \in \Sigma'^n} \Delta[n]_\sigma & \xrightarrow{q} & Sk^n Y \cup X
\end{array}
$$

Fig. 18

where i and f_{n-1} are inclusions, and q is induced by the singular simplices $\tilde{\sigma}\colon \Delta[n]_\sigma \to Y$. Hence the square of Fig. 19

$$\begin{array}{ccc} \coprod_{\sigma\in\Sigma'^n} \dot{\Delta}_\sigma & \longrightarrow & |Sk^{n-1}Y \cup X| \\ \downarrow{\scriptstyle |i|} & & \downarrow{\scriptstyle |f_{n-1}|} \\ \coprod_{\sigma\in\Sigma'^n} \Delta[n]_\sigma & \longrightarrow & |Sk^{n}Y \cup X| \end{array}$$

Fig. 19

is also cocartesian. But $|i|$, and therefore $|f_{n-1}|$ are closed injections. Passing to the direct limit, we see that

$$|f|\colon |X| = |Sk^{-1}Y \cup X| \to \varinjlim_n |Sk^n\, Y \cup X| = |X|$$

is a closed injection.

3.3. Let us show now that *the geometric realization functor commutes with kernels*: let then

$$N \xrightarrow{i} X \overset{f}{\underset{g}{\rightrightarrows}} Y$$

be an exact sequence of $\Delta^\circ\mathscr{E}$, N being a subcomplex of X, and i the inclusion. By 3.2, $|i|\colon |N| \to |X|$ is a closed injection, so that it is sufficient to show that $|i|$ induces a surjection of the set subordinated to $|N|$ onto the set of $x \in |X|$ such that $|f|(x) = |g|(x)$; but $|f|$ and $|g|$ are compatible with the operations of the group of automorphisms of $|\,?\,|$ in $|X|$ and $|Y|$, so that the equality $|f|(x) = |g|(x)$ implies that $|f|$ and $|g|$ coincide in the orbit of x, and hence that they coincide also in the closure of the cell C which contains x (§ 1.6). If $\sigma \in X_n$ is the non-degenerate simplex which defines the cell C and $\tilde{\sigma}\colon \Delta[n] \to X$ is the associated singular simplex, we see then that $|f\circ\tilde{\sigma}| = |g\circ\tilde{\sigma}|$; if $\tilde{\tau} \in Y_m$ is the non degenerate simplex which defines the cell $|f|(C) = |g|(C)$ and τ is the associated singular simplex, there are epimorphisms $p, q\colon \Delta[n] \to \Delta[m]$ of $\Delta^\circ\mathscr{E}$ such that $f\circ\tilde{\sigma} = \tilde{\tau}\circ p$ and $g\circ\tilde{\sigma} = \tilde{\tau}\circ q$; from the equality $|\tilde{\tau}\circ p| = |\tilde{\tau}\circ q|$ and from the fact that $|\tilde{\tau}|$ induces an injection of $\mathring{\Delta}^m$ into $|Y|$, it follows then that $|p|$ and $|q|$ coincide in the interiors of $|\Delta[n]|$, and hence in the whole of $|\Delta[n]|$. Since the equality $|p| = |q|$ clearly implies $p = q$, we have $f\circ\tilde{\sigma} = g\circ\tilde{\sigma}$, i.e. $f(\sigma) = g(\sigma)$ and $\sigma \in N$. It follows that x belongs to the image of $|i|$.

3.4. In order to prove that $|\,?\,|$ commutes with finite inverse limits, it remains to be shown that $|\,?|$ commutes with direct products. We will see first that, for each pair (p, q) of natural integers, the canonical map

$$|\Delta[p] \times \Delta[q]| \to |\Delta[p]| \times |\Delta[q]|$$

is a homeomorphism. We use for this the presentation of Fig. 20 given in II, 5.5.

$$\coprod_{1\leq i<j\leq \binom{p+q}{p}} \Delta[n_{c(i)\cap c(j)}] \underset{v}{\overset{u}{\rightrightarrows}} \coprod_{1\leq i\leq \binom{p+q}{p}} \Delta[n_{c(i)}] \xrightarrow{\pi} \Delta[p]\times\Delta[q].$$

Fig. 20

In this presentation, $c(i)$ runs through the maximal chains of the ordered set $[p]\times[q]$; the morphism π is defined by the inclusions of the chains of $c(i)$ into $[p]\times[q]$; the morphism u and v are defined by the inclusions of the chains $c(i)\cap c(j)$ into $c(i)$ and $c(j)$.

Recall that a (p, q)-shuffle is defined by two sequences of integers $i_1<i_2<\cdots<i_p$ and $j_1<j_2<\cdots<j_p$ such that $\{i_1, \ldots, i_p, j_1, \ldots, j_q\} = \{1, 2, \ldots, p+q\}$. With each maximal chain $c(i)$, we associate a (p, q)-shuffle in the following way (Fig. 21):

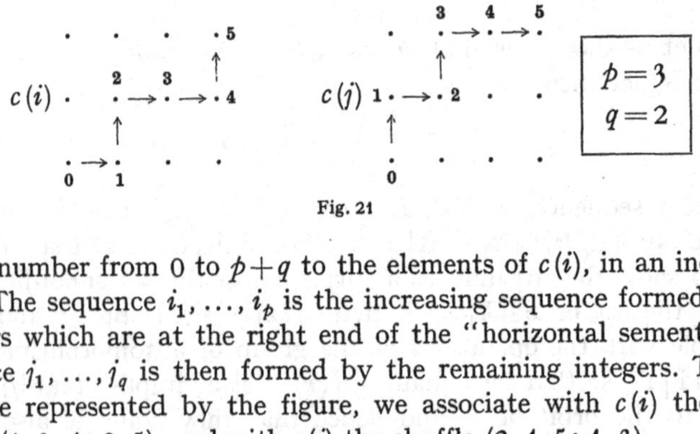

Fig. 21

Assign number from 0 to $p+q$ to the elements of $c(i)$, in an increasing order. The sequence $i_1, \ldots, i_p$ is the increasing sequence formed by the numbers which are at the right end of the "horizontal sements"; the sequence $j_1, \ldots, j_q$ is then formed by the remaining integers. Thus, in the case represented by the figure, we associate with $c(i)$ the (p, q)-shuffle $(1, 3, 4; 2, 5)$, and with $c(j)$ the shuffle $(2, 4, 5; 1, 3)$.

Let us write $s_0, s_1, \ldots, s_{p+1}$ for the components of a point of $\boldsymbol{R}^{p+2}$, $t_0, t_1, \ldots, t_{q+1}$ for those of a point of $\boldsymbol{R}^{q+2}$, and $u_0, u_1, \ldots, u_{p+q+1}$ for those of a point of $\Delta^{p+q}\subset\boldsymbol{R}^{p+q+2}$. We write f_i for the map from $\Delta^{p+q}\simeq |\Delta[n_{c(i)}]|$ to $\boldsymbol{R}^{p+2}\times\boldsymbol{R}^{q+2}$ which sends $u=(u_0, u_1, \ldots, u_{p+q+1})$ to the point of $\boldsymbol{R}^{p+2}\times\boldsymbol{R}^{q+2}$ defined by the equalities

$$s_0=u_0,\ s_1=u_{i_1},\ \ldots,\ s_p=u_{i_p},\ s_{p+1}=u_{p+q+1}$$

$$t_0=u_0,\ t_1=u_{j_1},\ \ldots,\ t_q=u_{j_q},\ t_{q+1}=u_{p+q+1}.$$

It is clear that f_i induces a homeomorphism from Δ^{p+q} onto the geometric simplex $\Delta_{c(i)}$ of $\boldsymbol{R}^{p+2}\times\boldsymbol{R}^{q+2}$ defined, in the case represented by the figure, by the relations

$$0=s_0=t_0\leq s_1\leq t_1\leq s_2\leq s_3\leq t_2\leq s_4=t_3=1.$$

If we compose f_i with the map $|\Delta[n_{c(i)\cap c(j)}]|\to|\Delta[n_{c(i)}]|$ defined by the inclusion of $c(i)\cap c(j)$ into $c(i)$, we obtain a homeomorphism $f_{i,j}$ from $|\Delta[n_{c(i)\cap c(j)}]|$ onto $\Delta_{c(i)}\cap\Delta_{c(j)}$. In the case of the figure, $\Delta_{c(j)}$ is defined by the relations

$$0 = s_0 = t_0 \leqq t_1 \leqq s_1 \leqq t_2 \leqq s_2 \leqq s_3 \leqq s_4 = t_3 = 1$$

and $\Delta_{c(i)} \cap \Delta_{c(j)}$ by the relations

$$0 = s_0 = t_0 \leqq s_1 = t_1 \leqq s_2 = s_3 = t_2 \leqq s_4 = t_3 = 1 .$$

In that case, $f_{i,j}$ is the map $(u_0, u_1, u_2, u_3) \rightsquigarrow (u_0, u_1, u_2, u_2, u_3; u_0, u_1, u_2, u_3)$.

To sum up, we obtain a commutative diagram (Fig. 22)

$$\begin{array}{ccccc} \coprod |\Delta[n_{c(i)\cap c(j)}]| & \overset{|u|}{\underset{|v|}{\rightrightarrows}} & \coprod |\Delta[n_{c(i)}]| & \xrightarrow{|\pi|} & |\Delta[p] \times \Delta[q]| \\ \downarrow \scriptstyle{\coprod f_{i,j}} & & \downarrow \scriptstyle{\coprod f_i} & & \\ \coprod (\Delta_{c(i)} \cap \Delta_{c(j)}) & \overset{u'}{\underset{v'}{\rightrightarrows}} & \coprod \Delta_{c(i)} & & \end{array}$$

Fig. 22

where the maps u' and v' are defined by the inclusion of $\Delta_{c(i)} \cap \Delta_{c(j)}$ into $\Delta_{c(i)}$ and $\Delta_{c(j)}$. The inclusion of $\Delta_{c(i)}$ into $\Delta^p \times \Delta^q \subset \boldsymbol{R}^{p+2} \times \boldsymbol{R}^{q+2}$ allow us to identify coker (u', v') with $\Delta^p \times \Delta^q$. Since $\coprod f_{i,j}$ and $\coprod f_i$ are homeomorphisms, and since the first row of the diagram is exact, the diagram induces an isomorphism from $|\Delta[p] \times \Delta[q]|$ onto $\Delta^p \times \Delta^q$; this isomorphism obviously coincides with the canonical map.

3.5. Now let X and Y be two arbitrary complexes. We know that X and Y are identified with the direct limits of the functors $d_X: \Delta/X \to \Delta^\circ \mathscr{E}$ and $d_Y: \Delta/Y \to \Delta^\circ \mathscr{E}$ (see II, 1.1). From this we deduce, as in II, 7.5 the following sequence of isomorphisms:

$$\begin{aligned} |X \times Y| &\simeq \left|\varinjlim_{\alpha} d_X(\alpha) \times \varinjlim_{\beta} d_Y(\beta)\right| \simeq \left|\varinjlim_{\alpha,\beta} d_X(\alpha) \times d_Y(\beta)\right| \\ &\simeq \varinjlim_{\alpha,\beta} |d_X(\alpha) \times d_Y(\beta)| \simeq \varinjlim_{\alpha,\beta} (|d_X(\alpha)| \times |d_Y(\beta)|) \\ &\simeq \left(\varinjlim_{\alpha} |d_X(\alpha)| \times \varinjlim_{\beta} |d_Y(\beta)|\right)_{\mathscr{K}e} \simeq (|X| \times |Y|)_{\mathscr{K}e}. \end{aligned}$$

This time, we used the fact that the product commutes with direct limits in the category of Kelley spaces (2.12). We have to see now that by putting the isomorphisms desribed above, one at the end of the other, we get the canonical map from $|X \times Y|$ into $(|X| \times |Y|)_{\mathscr{K}e}$. This is easy.

3.6. Finally, let us show that the geometric realization functor is conservative: let $f: X \to Y$ be a morphism of $\Delta^\circ \mathscr{E}$ which is not invertible. Let us show that $|f|$ is not invertible.

If f is a monomorphism, Y contains a non-degenerate simplex σ which is not the image of a simplex of X. Since the orbits of the group of automorphisms of $|?|$ in $|Y|$ are indexed by the non-degenerate simplices, the orbit of index σ cannot belong to the image of $|f|$.

If f is not a monomorphism, let us consider the equivalence relation defined by f: we consider then the subcomplex $X \underset{Y}{\times} X$ of $X \times X$ whose q-simplices (x, y) are such that $f_q(x) = f_q(y)$. Then $X \underset{Y}{\times} X$ contains the diagonal subcomplex D_X of $X \times X$ formed by all simplices (x, x); but $X \underset{Y}{\times} X$ is different from D_X. Since the geometric realization functor commutes with finite inverse limits, $X \underset{Y}{\times} X$ is identified with the equivalence relation $(|X| \underset{|Y|}{\times} |X|)_{\mathcal{K}e}$ defined by $|f|$ and, by above, this relation is different from the diagonal of $(|X| \times |X|)_{\mathcal{K}e}$, which is identified with $|D_X|$. Hence the canonical map from $|X|$ into the quotient $|X|/(|X| \underset{|Y|}{\times} |X|)_{\mathcal{K}e}$ is not injective; since $|f|$ factors through this quotient, $|f|$ is not injective.

3.7. Note that $|\,?\,|$ does not necessarily commute with infinite inverse limits: for example, for each cardinal number c, the image of the continuous map from $|\Delta[1]^c|$ into $|\Delta[1]|^c$ is the subset of $|\Delta[1]|^c$ formed by all points whose components take only a finite number of different values.

4. Geometric Realization of a Locally Trivial Morphism

4.1. We say that a *morphism* $f\colon Y \to X$ *of* $\Delta^\circ\mathcal{E}$ *is trivial* if there exists a complex F and an isomorphism $\alpha\colon X \times F \xrightarrow{\sim} Y$ such that $\mathrm{pr}_1 = f \circ \alpha$, where pr_1 is the canonical projection of $X \times F$ onto X. We say then that F, which is isomorphic to the complex $f^{-1}(x) = \Delta[0] \underset{\tilde{x}, f}{\times} Y$ for each singular simplex $\tilde{x}\colon \Delta[0] \to X$, is "the" *fibre* of f. We say that a morphism $f\colon Y \to X$ of $\Delta^\circ\mathcal{E}$ is *locally trivial* if, for each singular simplex $\tilde{\sigma}\colon \Delta[n] \to X$, the projection of the fibred product $\Delta[n] \underset{\tilde{\sigma}, f}{\times} Y$ onto $\Delta[n]$ is trivial. The fibre of this projection is then a complex F_σ; when all fibres F_σ are isomorphic to the same complex F, we say that f is locally trivial with fibre F; this happens, for instance, when X is connected (II, 7.3).

Similarly, we say that a ***morphism*** $\boldsymbol{u}$: $\boldsymbol{L} \to \boldsymbol{K}$ of *Kelley spaces is trivial*, with fibre $\boldsymbol{T}$, if there is an isomorphism $\beta\colon (\boldsymbol{K} \times \boldsymbol{T})_{\mathcal{K}e} \xrightarrow{\sim} \boldsymbol{L}$ such that $\mathrm{pr}_1 = \boldsymbol{u} \circ \beta$, where pr_1 always denotes the canonical projection of a product onto its first factor. We say that $\boldsymbol{u}$ is locally trivial if each point x of $\boldsymbol{K}$ has an open neighborhood $\boldsymbol{U}$ such that $\boldsymbol{u}$ induces a trivial morphism from $\boldsymbol{u}^{-1}(\boldsymbol{U})$ into $\boldsymbol{U}$ (these open subsets are Kelley spaces, by I, 1.5.3). If the fibres of there trivial morphisms are all isomorphic to the same (KELLEY) space $\boldsymbol{T}$, we say that $\boldsymbol{u}$ is locally trivial with fibre $\boldsymbol{T}$. In the present context, the products we consider are those of the category $\mathcal{K}e$, so that the definition given here does not respect the usual terminology.

4.2. *Theorem: The geometric realization functor* $|?|: \Delta^\circ\mathscr{E}\to\mathscr{K}e$ *transforms a locally trivial morphism with fibre* F *into a locally trivial morphism with fibre* $|F|$.

Let $f: Y\to X$ be a locally trivial morphism with fibre F of $\Delta^\circ\mathscr{E}$, x a point of $|X|$, m the dimension, and σ the index of the cell of $|X|$ which contains x (the notations are those of 1.7 and 1.8). Let $\boldsymbol{U}_m$ be the cell of x and $\boldsymbol{U}$ (resp. $\boldsymbol{U}_n$) be the extension of $\boldsymbol{U}_m$ to $|X|$ (resp. to $|Sk^nX|$, $n\geqq m$) of constant width 1 (with the notations of 1.8, $\boldsymbol{U}$ is then the open neighborhood $\boldsymbol{V}_x^{1,1}$ of x). We will see that $|f|$ induces a trivial morphism with fibre $|F|$ from $|f|^{-1}(\boldsymbol{U})$ to $\boldsymbol{U}$. We must then construct an isomorphism $\boldsymbol{\alpha}: (\boldsymbol{U}\times|F|)_{\mathscr{K}e}\to|f|^{-1}(\boldsymbol{U})$ such that $\mathrm{pr}_1=\boldsymbol{q}\circ\boldsymbol{\alpha}$, where $\boldsymbol{q}: |f|^{-1}(\boldsymbol{U})\to\boldsymbol{U}$ is the morphism induced by $|f|$.

Since $\boldsymbol{U}$ is the direct limit of the $\boldsymbol{U}_n$, since $(\boldsymbol{U}\times|F|)_{\mathscr{K}e}$ is the direct limit of the $(\boldsymbol{U}_n\times|F|)_{\mathscr{K}e}$ by 2.1, and since $|f|^{-1}(\boldsymbol{U})$ is the direct limit of the $|f|^{-1}(\boldsymbol{U}_n)$ by 2.2. ($|f|^{-1}(\boldsymbol{U})$ and $|f|^{-1}(\boldsymbol{U}_n)$ are identified with the fibred products $(\boldsymbol{U}\underset{|X|}{\times}|Y|)_{\mathscr{K}e}$ and $(\boldsymbol{U}_n\underset{|X|}{\times}|Y|)_{\mathscr{K}e}$, it is sufficient to construct a direct system of isomorphisms $\alpha_n: (\boldsymbol{U}_n\times|F|)_{\mathscr{K}e}\to|f|^{-1}(\boldsymbol{U}_n)$ such that $\mathrm{pr}_1=\boldsymbol{q}_n\circ\boldsymbol{\alpha}_n$, where $\boldsymbol{q}_n: |f|^{-1}(\boldsymbol{U}_n)\to\boldsymbol{U}_n$ is the morphism induced by $|f|$.

4.2.1. The construction of $\boldsymbol{\alpha}_m$ is quite simple, since, by hypothesis, there is an isomorphism β from $\Delta[m]\times F$ onto $\Delta[m]\underset{\tilde\sigma,f}{\times}Y$ which is compatible with canonical projections of these complexes onto $\Delta[m]$. Since $|\tilde\sigma|$ induces an isomorphism from the interior of $|\Delta[m]|$ onto $\boldsymbol{U}_m$, $(\boldsymbol{U}_m\times|F|)_{\mathscr{K}e}$ and $|f|^{-1}(\boldsymbol{U}_m)$ are identified with open sets of $|\Delta[m]\times F|$ and $|\Delta[m]\underset{\tilde\sigma,f}{\times}Y|$, by 3.1. We can then take for $\boldsymbol{\alpha}_m$ the isomorphism induced by $|\beta|$.

Suppose now that we have $\boldsymbol{\alpha}_{n-1}$, and let us construct $\boldsymbol{\alpha}_n$ $(n>m)$. In order to do this, consider the following commutative diagram of $\mathscr{K}e$ (Fig. 23).

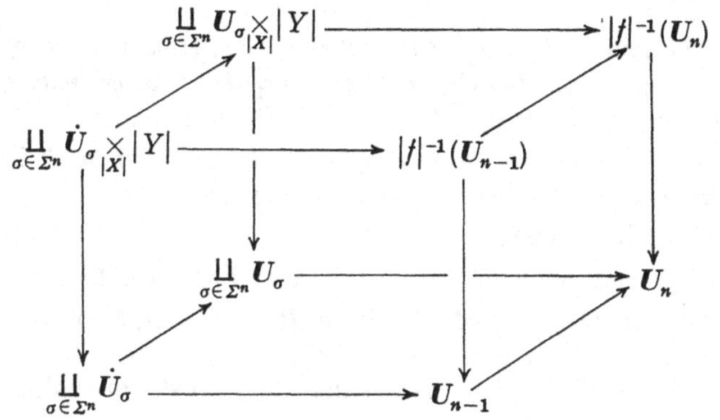

Fig. 23

The bottom square is that of 1.7 ($\dot{U}_\sigma$ is the space $|\tilde{\sigma}|^{-1}(U_{n-1})$). The top square is obtained from the bottom one by the change of base $|f|: |Y| \to |X|$. By 1.7, the bottom square is cocartesian; by 2.2 the same holds for the top one.

4.2.2. On the other hand, thei somorphism $\alpha_{n-1}: (U_{n-1} \times |F|)_{\mathscr{K}e} \simeq |f|^{-1}(U_{n-1})$ induces, by a change of base, isomorphisms $\alpha_\sigma: \dot{U}_\sigma \times |F| \to \dot{U}_\sigma \underset{|X|}{\times} |Y|$ which are compatible with the projections onto $\dot{U}_\sigma$ (see 2.1.1). Similarly, since f is locally trivial, there are isomorphisms $\beta_\sigma: \Delta[n]_\sigma \times F \simeq \Delta[n]_\sigma \underset{X}{\times} Y$, which are compatible with the projections onto $\Delta[n]_\sigma$. Since the geometric realization functor commutes with products and fibred products, $|\beta_\sigma|$ induces, by restriction, isomorphisms

$$\gamma_\sigma: U_\sigma \times |F| \simeq U_\sigma \underset{|X|}{\times} |X| \quad \text{and} \quad \delta_\sigma: \dot{U}_\sigma \times |F| \simeq \dot{U}_\sigma \underset{|X|}{\times} Y.$$

By the lemma below, the automorphism $\delta_\sigma^{-1} \circ \alpha_\sigma$, which is compatible with the canonical projection of $\dot{U}_\sigma \times |F|$ onto $\dot{U}_\sigma$, can be extended to an automorphism ψ_σ of $U_\sigma \times |F|$, which is compatible with the projection onto U_σ. It follows that we have a commutative diagram (Fig. 24):

$$\begin{array}{ccccc}
\coprod U_\sigma \times |F| & \longleftarrow & \coprod \dot{U}_\sigma \times |F| & \longrightarrow & (U_{n-1} \times |F|)_{\mathscr{K}e} \\
{\scriptstyle \coprod \gamma_\sigma \circ \psi_\sigma}\downarrow & & {\scriptstyle \coprod \alpha_\sigma}\downarrow & & {\scriptstyle \alpha_{n-1}}\downarrow \\
\coprod U_\sigma \underset{|X|}{\times} |Y| & \longleftarrow & \coprod \dot{U}_\sigma \underset{|X|}{\times} |Y| & \longrightarrow & |f|^{-1}(U_{n-1})
\end{array}$$

Fig. 24

where the horizontal arrows are obvious; moreover, this diagram is compatible with the canonical projections onto the bases. The "vertical" isomorphisms induce an isomorphism α_n from the amalgamated sum $(U_n \times |F|)_{\mathscr{K}e}$ of the upper part of the diagram onto the amalgamated sum $|f|^{-1}(U_n) \simeq (U_n \underset{|X|}{\times} |Y|)_{\mathscr{K}e}$ of the lower part (see 2.2). This completes the construction of α_n.

4.2.3. *Lemma: Let* A *and* F *be Kelley spaces,* B *be a retract of* A *and* φ *an automorphism of* $(B \times F)_{\mathscr{K}e}$ *with* $\mathrm{pr}_1 = \mathrm{pr}_1 \circ \varphi$. *Then* φ *can be extended to an automorphism* ψ *of* $(A \times F)_{\mathscr{K}e}$ *such that* $\mathrm{pr}_1 = \mathrm{pr}_1 \circ \psi$.

The map φ is of the form $(x, y) \rightsquigarrow (x, \Phi(x) \cdot y)$, where Φ is a map from B to $\mathscr{T}\!op(F, F)$ satisfying the following conditions:

a) $\Phi(x)$ is invertible for all $x \in B$.

b) Φ is a continuous map from B to $CO(F, F)$ (see 2.1.2).

c) The map $\Phi^{-1}: x \rightsquigarrow \Phi(x)^{-1}$ from B to $CO(F, F)$ is continuous (the inverse of φ is the continuous map $(x, y) \rightsquigarrow (x, \Phi(x)^{-1} \cdot y)$. If r is a retraction of A onto B, it is clear that the map $\Psi = \Phi \circ r$ also satisfies a), b), c) "mutatis mutandis". We can then write $\psi(x, y) = (x, \Psi(x) \cdot y)$.

Chapter Four

The Homotopic Category

1. Homotopies

1.1. Let Y and Z be simplicial sets, and let $\mathcal{H}om(Y, Z)$ be the complex defined in II, 2.5.3. A vertex of $\mathcal{H}om(Y, Z)$ is a morphism $f: Y \to Z$. If f and g are two vertices, a 1-simplex h of $\mathcal{H}om(Y, Z)$ such that $d_1 h = f$ and $d_0 h = g$ is a morphism $h: \Delta[1] \times Y \to Z$ which makes the following triangles commutative (Fig. 25).

$$\begin{array}{ccc} \Delta[0]\times Y \simeq Y & & \\ {\scriptstyle \Delta(\partial_1^1)\times Y}\swarrow & \searrow{\scriptstyle f} & \\ \Delta[1]\times Y & \xrightarrow{h} & Z \end{array} \qquad \begin{array}{ccc} \Delta[0]\times Y \simeq Y & & \\ {\scriptstyle \Delta(\partial_1^0)\times Y}\swarrow & \searrow{\scriptstyle g} & \\ \Delta[1]\times Y & \xrightarrow{h} & Z \end{array}$$

Fig. 25

We will say simply that h is a *homotopy connecting* f with g.

More generally, let us use the complex I_n of II, 2.5.1 and II, 5.2.1. We will say that a morphism $h: I_n \times Y \to Z$ is a *composed homotopy* connecting f with g when the triangles of Fig. 26

$$\begin{array}{ccc} \Delta[0]\times Y \simeq Y & & \\ {\scriptstyle \varepsilon_0\times Y}\swarrow & \searrow{\scriptstyle f} & \\ I_n\times Y & \xrightarrow{h} & Z \end{array} \qquad \begin{array}{ccc} \Delta[0]\times Y \simeq Y & & \\ {\scriptstyle \varepsilon_n\times Y}\swarrow & \searrow{\scriptstyle g} & \\ I_n\times Y & \xrightarrow{h} & Z \end{array}$$

Fig. 26

are commutative.

Finally, we will say, that two morphisms f and g from Y to Z are *homotopic* if there is an integer n and a composed homotopy of length n connecting f with g. The homotopy relation is obviously an equivalence relation. Take in particular $Y = \Delta[0]$: $\mathcal{H}om(\Delta[0], Z)$ is then identified with Z and the homotopy equivalence classes are simply *the connected components* of Z (see II, 7.3); the set of these classes is denoted by $\Pi_0 Z$. Going back to the general case, it is clear that two morphisms f and g from Y to Z are homotopic if and only if they belong to the same connected component of $\mathcal{H}om(X, Y)$.

1.2. An example: Order the set $\{0, 1, 2\}$ by the relation defined by the inequalities $0 \leqq 1 \geqq 2$. Let r_n^i be the increasing map from $\{0, 1, 2\} \times [n]$

to $[n]$ defined by the following formulas:

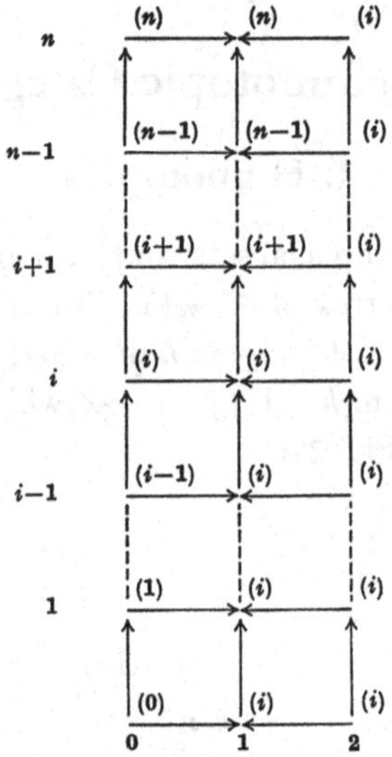

Fig. 27

$r_n^i(0,j)=j$, $r_n^i(1,j)=j$ if $j\geqq i$, $r_n^i(1,j)=i$, if $j\leqq i$ and $r_n^i(2,j)=i$ (in the Fig. 27, we wrote beside each point of $\{0, 1, 2\}\times[2]$ the value taken by r_n^i).

By II.5, the complex associated with the ordered set $\{0, 1, 2\}\times[n]$ is identified with $I_2\times\Delta[n]$. Hence the map r_n^i induces a morphism h: $I_2\times\Delta[n]\to\Delta[n]$; this morphism is a composed homotopy of length 2 connecting the identity of $\Delta[n]$ with the "projection of $\Delta[n]$ onto the i-th vertex".

Note here that if $i\neq n$, there is no homotopy connecting the identity of $\Delta[n]$ with the "projection of $\Delta[n]$ onto the i-th vertex".

1.3. If Y and Z are two complexes, we will write temporarily $\Pi_0(Y,Z)$ for the set of equivalence classes of $\Delta^\circ\mathscr{E}(Y,Z)$ under homotopy. Hence this set is also identified with the set of connected components of the complex $\mathscr{H}om(Y,Z)$ (1.1). We will see that the *composition maps*

$$\Delta^\circ\mathscr{E}(X,Y)\times\Delta^\circ\mathscr{E}(Y,Z)\to\Delta^\circ\mathscr{E}(X,Z)$$

of $\Delta^\circ\mathscr{E}$ are compatible with the homotopy relation:

In order to do this, let us associate with each morphism h: $\Delta[n]\times X\to Y$ the morphism $\tilde{h}$: $\Delta[n]\times X\to\Delta[n]\times Y$ whose components are the canonical projection of $\Delta[n]\times X$ onto $\Delta[n]$ on the one hand, and the morphism h on the other. We can then defined a morphism of complexes

$$\nu_{X,Y,Z}\colon \mathscr{H}om(X,Y)\times\mathscr{H}om(Y,Z)\to\mathscr{H}om(X,Z)$$

by sending the pair (h, k) belonging to $\mathcal{H}om(X, Y)_n \times \mathcal{H}om(Y, Z)_n$ to the morphism $l\colon \Delta[n] \times X \to Z$ defined by the equation $\tilde{l} = \tilde{k} \circ \tilde{h}$. In other words, l is the composition

$$\Delta[n] \times X \xrightarrow{\tilde{h}} \Delta[n] \times Y \xrightarrow{k} Z.$$

The map induced by $\nu_{X,Y,Z}$ on the 0-simplices is simply the composition map. On the other hand, since the set $\Pi_0(A \times B)$ of connected components of a product is identified with the product $\Pi_0(A) \times \Pi_0(B)$, $\nu_{X,Y,Z}$ induces a map

$$\mu_{X,Y,Z}\colon \Pi_0(X, Y) \times \Pi_0(Y, Z) \to \Pi_0(X, Z)$$

on the connected components; this map is obviously compatible with the canonical surjection from the 0-simplices onto the set of connected components.

The maps $\mu_{X,Y,Z}$ are the composition maps of a category whose objects are the simplicial sets, and whose morphisms are the sets $\Pi_0(X, Y)$. This category will be called *the category of complexes modulo homotopy*, and will be written $\overline{\Delta^\circ\mathcal{E}}$. We will call *canonical functor from* $\Delta^\circ\mathcal{E}$ *to* $\overline{\Delta^\circ\mathcal{E}}$ the functor which induces the identity on the set of objects and the canonical maps from $\Delta^\circ\mathcal{E}(X, Y)$ to $\Pi_0(X, Y)$ on the sets of morphisms.

1.4. *Proposition: The canonical functor from* $\Delta^\circ\mathcal{E}$ *to* $\overline{\Delta^\circ\mathcal{E}}$ *commutes with finite sums and products.*

Let $(Y_i)_{i\in I}$ be a finite family of complexes and let pr_i be the projection of the product $\prod_{i\in I} Y_i$ onto the i-th factor. For each integer n, the pr_i induce a bijection

$$\psi_n\colon \Delta^\circ\mathcal{E}\Big(\Delta[n] \times X, \prod_{i\in I} Y_i\Big) \xrightarrow{\sim} \prod_{i\in I} \Delta^\circ\mathcal{E}(\Delta[n] \times X, Y_i)$$

and hence an isomorphism

$$\psi\colon \mathcal{H}om\Big(X, \prod_{i\in I} Y_i\Big) \xrightarrow{\sim} \prod_{i\in I} \mathcal{H}om(X, Y_i).$$

Since the functor Π_0 commutes with finite products, the set of connected of the product $\prod_i \mathcal{H}om(X, Y_i)$ is identified with the product of the $\Pi_0(\mathcal{H}om(X, Y_i)) = \Pi_0(X, Y_i)$. Consequently, ψ induces on the connected components the isomorphism we looked for:

$$\Pi_0(\psi)\colon \Pi_0\Big(X, \prod_{i\in I} Y_i\Big) \xrightarrow{\sim} \prod_{i\in I} \Pi_0(X, Y_i).$$

For finite sums, the argument is similar.

1.5. Consider now three complexes X, Y and Z. By II, 2.5.3, we have, for each integer n, canonical bijections

$$\varphi_n\colon \Delta^\circ\mathcal{E}(\Delta[n] \times X \times Y, Z) \xrightarrow{\sim} \Delta^\circ\mathcal{E}(\Delta[n] \times X, \mathcal{H}om(Y, Z))$$

and hence functor isomorphisms

$$\Phi: \mathcal{H}om(X \times Y, Z) \simeq \mathcal{H}om(X, \mathcal{H}om(Y, Z))$$

and

$$\Pi_0(\Phi) = \Pi_0(X \times Y, Z) \rightarrow \Pi_0(X, \mathcal{H}om(Y, Z)).$$

This last isomorphism implies that if f and g are two homotopic morphisms from Z to Z', then $\mathcal{H}om(Y, f)$ and $\mathcal{H}om(Y, g)$ are homotopic; indeed it is sufficient to check that $\Pi_0(X, \mathcal{H}om(Y, f))$ and $\Pi_0(X, \mathcal{H}om(Y, g))$ coincide for all X; this follows obviously from the equality $\Pi_0(X \times Y, f) = \Pi_0(X \times Y, g)$. Similarly, we could see that $\mathcal{H}om(f', Z)$ and $\mathcal{H}om(g', Z)$ are homotopic when f' and g' are two homotopic morphisms from Y to Y'. In other words, *the functor*

$$\mathcal{H}om: (\Delta^\circ\mathcal{E})^\circ \times (\Delta^\circ\mathcal{E}) \rightarrow \Delta^\circ\mathcal{E}$$

defines, "by passing to the quotient", a functor

$$(\overline{\Delta^\circ\mathcal{E}})^\circ \times (\overline{\Delta^\circ\mathcal{E}}) \rightarrow \overline{\Delta^\circ\mathcal{E}}$$

which will still be written $\mathcal{H}om$, by an extension of notation.

2. Anodyne Extensions

From now on, for each natural integer $n \geqq 1$ and each natural integer $k \leqq n$, we will write $\Lambda^k[n]$ for the subcomplex of $\Delta[n]$ such that $\Lambda^k[n]_p$ is formed by the increasing maps $f: [p] \rightarrow [n]$ whose image does not contain the set $[n] - \{k\}$. The geometric realization of the complex $\Lambda^k[n]$ is the union of $n-1$ faces of the geometric simplex Δ^n, namely those which contain the k-th vertex of Δ^n. That is why we will say that $\Lambda^k[n]$ *is the* k-th *horn* of $\Delta[n]$.

2.1. A set A of monomorphisms of $\Delta^\circ\mathcal{E}$ is called *saturated* if it satisfies the following four conditions:

(i) All isomorphisms belong to A.

(ii) If the commutative square

$$\begin{array}{ccc} X & \longrightarrow & Y \\ {\scriptstyle \xi}\downarrow & & \downarrow{\scriptstyle \eta} \\ X' & \longrightarrow & Y' \end{array}$$

is cocartesian and if ξ belongs to A, then η belongs to A (i.e. A is stable under *push out*).

(iii) If there exists a commutative diagram

$$\begin{array}{ccccc} X & \xrightarrow{u} & Y & \xrightarrow{v} & X \\ {\scriptstyle \xi}\downarrow & & {\scriptstyle \eta}\downarrow & & \downarrow{\scriptstyle \xi} \\ X' & \xrightarrow{u'} & Y' & \xrightarrow{v'} & X' \end{array}$$

such that $v \circ u = \mathrm{Id}\, X$, $v' \circ u' = \mathrm{Id}\, X'$, and if η belongs to A, then ξ belongs to A (each *retract* of a morphism of A belongs to A).

(iv) A is stable under countable compositions and arbitrary direct sums: if $f_i\colon X_i \to X_{i+1}$ $(i=1, 2, \ldots)$ is a morphism of A, then $\mathrm{in}_1\colon X_1 \to \varinjlim_i X_i$ is a morphism of A; if $(g_\alpha\colon X_\alpha \to Y_\alpha)$ is a family of morphisms of A, then $\coprod_\alpha g_\alpha\colon \coprod_\alpha X_\alpha \to \coprod_\alpha Y_\alpha$ is a morphism of A.

The intersection of all saturated sets containing a given set of monomorphisms B, is saturated; we will call it the *saturated set generated by* B. In what follows, B will be one of the sets B_1, B_2, B_3 defined below:

B_1 is the set of all indusions of $\Lambda^k[n]$ into $\Delta[n]$, $n, k \in \mathbb{N}$, $n \geqq 1$, $k \leqq n$.

B_2 is the set of all inclusions of $\Delta[1] \times \dot{\Delta}[n] \cup \{e\} \times \Delta[n]$ into $\Delta[1] \times \Delta[n]$, $n \in \mathbb{N}$, $e = 0, 1$[1] ($\dot{\Delta}[n]$ is the boundary of $\Delta[n]$, II, 3.6).

B_3 is the set of all indusions of $\Delta[1] \times Y \cup \{e\} \times X$ into $\Delta[1] \times X$ (X runs through the simplicial sets, Y through the subcomplexes of X, and e through the numbers 0 and 1).

Theorem: The saturated sets generated by B_1, B_2, and B_3 coincide.

Let A_1, A_2 and A_3 be the saturated sets generated respectively by B_1, B_2 and B_3. We will show successively that $A_2 \subset A_1$, $A_3 \subset A_2$ and $A_1 \subset A_3$.

2.1.1. Suppose for instance that $e=1$, and let us show that the inclusion of $\Delta[1] \times \Delta[n] \cup \{e\} \times \Delta[n]$ into $\Delta[1] \times \Delta[n]$ belongs to A_1: let $c(i)\colon [n+1] \to [1] \times [n]$ be the strict increasing map which takes the values $(0, i)$ and $(1, i)$, and let $C_i\colon \Delta[n+1] \to \Delta[1] \times \Delta[n]$ be the morphism associated with $c(i)$ by the functor $C\colon \mathcal{O}r \to \Delta^\circ\mathcal{E}$ of II, 5.1. On the other hand, let D_{-1} be the complex $\Delta[1] \times \Delta[n] \cup \{e\} \times \Delta[n]$, and let D_i be the union of D_{-1} with the images of $C_0, C_1, \ldots, C_i$ $(0 \leqq i \leqq n)$. In the case, for instance, where $n=1$, the D_i have the following geometric realizations (Fig. 28)

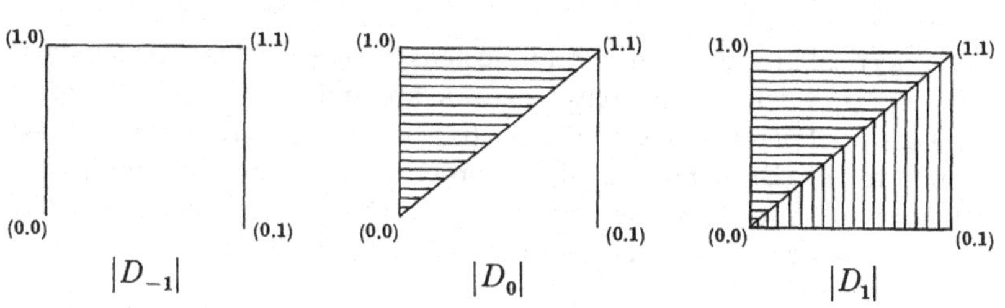

Fig. 28

[1] By an extension of notation $\{0\}$ and $\{1\}$ denote respectively the images of $\Delta[0]$ under the morphisms $\Delta(\partial_0^1)$ and $\Delta(\partial_0^0)$ from $\Delta[0]$ into $\Delta[1]$.

In the general case, D_n is simply $\Delta[1]\times\Delta[n]$ (see for example II, 5.5), so that it is sufficient to prove that the inclusions of D_{i-1} into D_i belong to $A_1 (0\leqq i\leqq n)$. This follows from condition (ii) above, and from the fact that the inverse image of D_{i-1} in $\Delta[n]$ under the morphism C_i is $\Lambda^{i+1}[n+1]$. Hence we have a cocartesian square (Fig. 29)

$$\begin{array}{ccc} \Lambda^{i+1}[n+1] & \xrightarrow{u} & D_{i-1} \\ \downarrow & & \downarrow \\ \Delta[n+1] & \xrightarrow{v} & D_i \end{array}$$

Fig. 29

where u and v are induced by C_i, the vertical arrows being inclusions.

2.1.2. Let us show now that, for each complex X and each subcomplex Y, the indusion of $\Delta[1]\times Y\cup\{e\}\times X$ into $\Delta[1]\times X$ belongs to A_2. As we have noticed in III, 3.2, the argument used in II, 3.8 shows that, if Σ'^n is the set of non-degenerate n-simplices of X which *do not belong to* Y, the square of Fig. 30

$$\begin{array}{ccc} \coprod_{\sigma\in\Sigma'^n} \dot{\Delta}[n]_\sigma & \longrightarrow & Y\cup Sk^{n-1}X \\ \downarrow & & \downarrow \\ \coprod_{\sigma\in\Sigma'^n} \Delta[n]_\sigma & \longrightarrow & Y\cup Sk^n X \end{array}$$

Fig. 30

is cocartesian. Consider then the commutative square of Fig. 31

$$\begin{array}{ccc} \coprod_{\sigma\in\Sigma'^n} (\Delta[1]\times\dot{\Delta}[n]_\sigma\cup\{e\}\times\Delta[n]_\sigma) & \longrightarrow & \Delta[1]\times(Y\cup Sk^{n-1}X)\cup\{e\}\times X \\ \downarrow & & \downarrow \\ \coprod_{\sigma\in\Sigma'^n} (\Delta[1]\times\Delta[n]_\sigma) & \longrightarrow & \Delta[1]\times(Y\cup Sk^n X)\cup\{e\}\times X \end{array}$$

Fig. 31

where vertical arrows are natural indusions, where $\dot{\Delta}[n]$ is the boundary of $\Delta[n]$, and where horizontal arrows are induced by the singular simplices $\tilde{\sigma}\colon \Delta[n]\to X$. By condition (iv) of 2.1, the left vertical arrow belongs to A_2. By condition (ii), in order to complete the proof, it is sufficient to show that the square in cocartesian; for then the right vertical arrow belongs to A_2, and hence also the inclusion of

$$\Delta[1]\times Y\cup\{e\}\times X=\Delta[1]\times(Y\cup Sk^{-1}X)\cup(e\times X)$$

into $\Delta[1]\times X=\varinjlim_n(\Delta[1]\times(Y\cup Sk^n X)\cup\{e\}\times X)$ (condition iv of 2.1).

In order to do this, let us consider the commutative diagram (Fig. 32)

$$\begin{array}{ccccc}
\{e\}\times X & \longleftarrow & \{e\}\times(Y\cup Sk^{n-1}X) & \longrightarrow & \Delta[1]\times(Y\cup Sk^{n-1}X)\\
\uparrow & & \uparrow & & \uparrow\\
\coprod_{\sigma\in\Sigma'^n}\{e\}\times\Delta[n]_\sigma & \longleftarrow & \coprod_{\sigma\in\Sigma'^n}\{e\}\times\dot{\Delta}[n]_\sigma & \longrightarrow & \coprod_{\sigma\in\Sigma'^n}\Delta[1]\times\dot{\Delta}[n]_\sigma\\
\downarrow & & \downarrow & & \downarrow\\
\coprod_{\sigma\in\Sigma'^n}\{e\}\times\Delta[n]_\sigma & \longleftarrow & \coprod_{\sigma\in\Sigma'^n}\{e\}\times\Delta[n]_\sigma & \longrightarrow & \coprod_{\sigma\in\Sigma'^n}\Delta[1]\times\Delta[n]_\sigma
\end{array}$$

Fig. 32

where all morphisms are the obvious ones.

It is clear that the amalgamated sums of the rows of this diagram are respectively $\Delta[1]\times(Y\cup Sk^{n-1}X)\cup\{e\}\times X$,

$$\coprod_{\sigma\in\Sigma'^n}(\Delta[1]\times\dot{\Delta}[n]_\sigma\cup\{e\}\times\Delta[n]_\sigma), \quad \text{and} \quad \coprod_{\sigma\in\Sigma'^n}(\Delta[1]\times\Delta[n]_\sigma);$$

similarly, the amalgamated sums of the columns are respectively $\{e\}\times X$, $\{e\}\times(Y\cup Sk^nX)$ and $\Delta[1]\times(Y\cup Sk^nX)$. Moreover, the amalgamated sum of the amalgamated sums of the columns is $\Delta[1]\times(Y\cup Sk^nX)\cup\{e\}\times X$. Since we can compute the direct limit of the diagram "row by row" or "column by column", $\Delta[1]\times(Y\cup Sk^nX)\cup\{e\}\times X$ is also the amalgamated sum of $\coprod_{\sigma\in\Sigma'^n}\Delta[1]\times\Delta[n]_\sigma$ and $\Delta[1]\times(Y\cup Sk^{n-1}X)\cup\{e\}\times X$ under $\coprod_{\sigma\in\Sigma'^n}(\Delta[1]\times\dot{\Delta}[n]_\sigma\cup\{e\}\times\Delta[n]_\sigma)$. Q.E.D.

2.1.3. It remains to be seen that the inclusions $\Lambda^k[n]\to\Delta[n]$ belong to A_3. Let u_n^k be the map $i\rightsquigarrow(1,i)$ from $[n]$ to $[1]\times[n]$ and v_n^k: $[1]\times[n]\to[n]$ be the retraction of u_n^k such that $v_n^k(1,i)=i$, $v_n^k(0,i)=i$ if $i\leq k$ and $v_n^k(0,i)=k$ if $i\geq k$. We see easily that the morphism $C(u_n^k)$: $\Delta[n]\to\Delta[1]\times\Delta[n]$, associated with u_n^k by the functor C of II.5, induces a morphism from $\Lambda^k[n]$ to $\Delta[1]\times\Lambda^k[n]\cup\{0\}\times\Delta[n]$. Similarly, $C(v_n^k)$: $\Delta[1]\times\Delta[n]\to\Delta[n]$ induces a morphism from $\Delta[1]\times\Lambda^k[n]\cup\{0\}\times\Delta[n]$ to $\Lambda^k[n]$ if $k<n$. We then have a commutative diagram (Fig. 33)

$$\begin{array}{ccccc}
\Lambda^k[n] & \longrightarrow & \Delta[n]\times\Lambda^k[n]\cup\{0\}\times\Delta[n] & \longrightarrow & \Lambda^k[n]\\
\downarrow & & \downarrow & & \downarrow\\
\Delta[n] & \xrightarrow{C(u_n^k)} & \Delta[1]\times\Delta[n] & \xrightarrow{C(v_n^k)} & \Delta[n]
\end{array}$$

Fig. 33

so that for $k<n$, the inclusion of $\Lambda^k[n]$ into $\Delta[n]$ belongs to A_3 by condition (iii) of 2.1.

For $k=n$ (and more generally for $k>0$), we can use in a similar way the map w_n^k: $i\rightsquigarrow(0,i)$ and retraction t_n^k of w_n^k such that $t_n^k(0,i)=i$, $t_n^k(1,i)=k$ if $i\leq k$ and $t_n^k(1,i)=i$ if $i\geq k$.

2.1.4. *Definition:* We call *anodyne extension* any monomorphism of $\Delta^\circ\mathscr{E}$ which belongs to the saturated set generated by B_1, B_2 or B_3.

2.2. *Proposition: If K is a subcomplex of L such that the inclusion of K is an anodyne extension and if Y is a subcomplex of X, then the inclusion*

$$K\times X\cup L\times Y\to L\times X$$

is an anodyne extension.

Let A be the set of monomorphisms $u\colon K'\to L'$ of $\Delta^\circ\mathscr{E}$ such that the morphism

$$K'\times X \coprod^{K'\times Y} L'\times Y\to L'\times X$$

induced by u is an anodyne extension. It is clear that A is a saturated set (2.1), so that it is sufficient to check that A contains B_3: let then Y' be a subcomplex of X' and let $K=\Delta[1]\times Y'\cup\{e\}\times X'$ and $L=\Delta[1]\times X$.

We then have

$$K\times X\cup L\times Y=\Delta[1]\times(Y'\times X\cup X'\times Y)\cup\{e\}\times X'\times X$$

and

$$L\times X=\Delta[1]\times X'\times X,$$

so that the inclusion of $K\times X\cup L\times Y$ into $L\times X$ belongs to B_3 and that the inclusion of K into L belongs to A.

2.3. By an extension of the definition, we will also call anodyne extension the image on an anodyne extension in the category $\overline{\Delta^\circ\mathscr{E}}$ of complexes modulo homotopy.

Theorem: The set of anodyne extensions of the category $\overline{\Delta^\circ\mathscr{E}}$ of complexes modulo homotopy admits a calculus of left fractions (I, 2.2).

Let A be the set of anodyne extensions of $\Delta^\circ\mathscr{E}$ and $\bar{A}$ the image of A in $\overline{\Delta^\circ\mathscr{E}}$. We must vertify that $\bar{A}$ satisfies conditions a), b), c) and d) of I, 2.2: conditions a) and b) of 2.2 follow from conditions (i) and (iv) of 2.1; similarly, c) follows from (ii); it remains then to prove d).

Let $s\colon X'\to X$ be an anodyne extension of $\Delta^\circ\mathscr{E}$ and $f, g\colon X\rightrightarrows Y$ two morphisms of $\Delta^\circ\mathscr{E}$ such that $f\circ s$ is homotopic to $g\circ s$. We must prove the existence of an anodyne extension $t\colon Y\to Y'$ of $\Delta^\circ\mathscr{E}$ such that $t\circ f$ is homotopic to $t\circ g$: let $h'\colon I_n\times X'\to Y$ be a composed homotopy connecting $f\circ s$ with $g\circ s$ (1.1) and let $\{0\}$ and $\{n\}$ denote the images of $\Delta[0]$ under the morphisms ε_0 and ε_n from $\Delta[0]$ to I_n. If we identify X' to a subcomplex of X by means of s, the inclusion of $\{0\}\times X\cup\{n\}\times X\cup I_n\times X'$ into $I_n\times X$ is an anodyne extension by 2.2. On the other hand, there is a morphism k from $\{0\}\times X\cup\{n\}\times X\cup I_n\times X'$ to Y which induces f on

$X \simeq \{0\} \times X$, g on $X \simeq \{n\} \times X$, and h' on $I_n \times X'$. We then have a diagram (Fig. 34)

$$\begin{array}{c} \{0\} \times X \cup \{n\} \times X \cup I_n \times X' \xrightarrow{k} Y \\ \big\downarrow \text{incl.} \\ I_n \times X \end{array}$$

Fig. 34

Let Y' be the amalgamated sum of this diagram, and t and h the canonical morphisms from Y and $I_n \times X$ to the amalgamated sum Y'. By (ii) 2.1, t is an anodyne extension; since h is a composed homotopy connecting $t \circ f$ with $t \circ g$, the theorem is proved.

2.3.1. *Definition: We will call homotopic category, the category of fractions of $\overline{\Delta^\circ \mathscr{E}}$ for the set of anodyne extensions. This category will be denoted by $\mathscr{H}$.*

By proposition I, 3.1, the canonical functor $P_{\bar{A}}$: $\overline{\Delta^\circ \mathscr{E}} \to \mathscr{H}$ commutes with finite direct limits, and hence in particular with finite direct sums. Proposition 1.4 implies that finite direct sums exist in $\mathscr{H}$ and can be constructed as in $\Delta^\circ \mathscr{E}$. We will see later that the same holds for direct products.

3. Kan Complexes

In this paragraph, we intend to show that the canonical functor $P_{\bar{A}}$: $\overline{\Delta^\circ \mathscr{E}} \to \mathscr{H}$ has a right adjoint; by I, 4.1, we will then be in the situation of proposition I, 1.3. According to I, 4.1, it is sufficient to associate with each complex X an anodyne extension $a(X)$: $X \to X_K$ such that X_K is left closed for $\bar{A}$. In order to do this, we will study first a certain class of complexes which are left closed for $\bar{A}$.

3.1. *Definition:* A morphism p: $E \to B$ of $\Delta^\circ \mathscr{E}$ *is a fibration in the sense of* Kan (or simply a *fibration*) if, for each anodyne extension i: $K \to L$ and each commutative square

$$(*) \qquad \begin{array}{ccc} K & \xrightarrow{u} & E \\ {\scriptstyle i}\big\downarrow & & \big\downarrow{\scriptstyle p} \\ L & \xrightarrow{v} & B \end{array}$$

there is a morphism w: $L \to E$ of $\Delta^\circ \mathscr{E}$ such that $u = w \circ i$ and $v = p \circ w$. A simplicial set X is a Kan *complex* if the unique morphism $X \to \Delta[0]$ is a fibration in the sense of Kan.

Let p: $E \to B$ be a morphism of $\Delta^\circ \mathscr{E}$ and P the set of monomorphisms i: $K \to L$ of $\Delta^\circ \mathscr{E}$ such that, for each commutative square of the form $(*)$, there is a morphism w satisfying the equalities $u = w \circ i$ and $c = p \circ w$. It is clear that P is saturated in the sense of 2.1. In order to prove that p is a fibration, it will then be sufficient to prove that P contains one of the sets B_1, B_2 or B_3 of 2.1.

3.1.1. The following properties of fibrations follow directly from the definitions:

(i bis) Every isomorphism is a fibration in the sense of KAN.

(ii bis) If the commutative square

$$\begin{array}{ccc} E' & \longrightarrow & E \\ {\scriptstyle p'}\downarrow & & \downarrow{\scriptstyle p} \\ B' & \longrightarrow & B \end{array}$$

is cartesian and if p is a fibration, then p' is a fibration.

(iii bis) If there is a commutative diagram

$$\begin{array}{ccccc} E & \xrightarrow{u} & E' & \xrightarrow{v} & E \\ {\scriptstyle p}\downarrow & & {\scriptstyle p'}\downarrow & & {\scriptstyle p}\downarrow \\ B & \xrightarrow{u'} & B' & \xrightarrow{v'} & B \end{array}$$

such that $v \circ u = \mathrm{Id}\, E$, $v' \circ u' = \mathrm{Id}\, B$, and if p' is a fibration in the sense of KAN, then p is a fibration.

(iv bis) consider an infinite countable sequence

$$\ldots X_3 \xrightarrow{p_3} X_2 \xrightarrow{p_2} X_1 \xrightarrow{p_1} X_0$$

such that all p_n are fibrations. Then the canonical projection $p: \varprojlim_n X_n \to X_0$ is a fibration; if $(p_\alpha: E_\alpha \to B_\alpha)$ is a family of fibrations, the same holds for $\prod_\alpha p_\alpha: \prod_\alpha E_\alpha \to \prod_\alpha B_\alpha$.

Note in particular that the product $X \times Y$ of two Kan complexes is a Kan complex.

3.1.2. Here are other useful properties of fibrations in the sense of KAN:

If $p: E \to B$ is a fibration, $\mathcal{H}om(X, p): \mathcal{H}om(X, E) \to \mathcal{H}om(X, B)$ is a fibration for each simplicial set X.

Consider a commutative square (Fig. 35)

$$\begin{array}{ccc} K & \xrightarrow{u} & \mathcal{H}om(X, E) \\ {\scriptstyle i}\downarrow & & \downarrow{\scriptstyle \mathcal{H}om(X, p)} \\ L & \xrightarrow{v} & \mathcal{H}om(X, B) \end{array}$$

Fig. 35

where i is an anodyne extension. We have seen (II, 2.5.3) that the functor $? \times X$ is left adjoint to the functor $\mathcal{H}om(X, ?)$: then the morphisms $u': K \times X \to E$ and $v': L \times X \to B$ associated with u and v by the functor isomorphism

$$\Delta^\circ\mathcal{E}(? \times X, ?) \simeq \Delta^\circ\mathcal{E}(?, \mathcal{H}om(X, ?))$$

make the following square commutative

$$\begin{array}{ccc} K\times X & \xrightarrow{u'} & E \\ {\scriptstyle i\times X}\downarrow & & \downarrow{\scriptstyle p} \\ L\times X & \xrightarrow{v'} & B \end{array}$$

Since i is an anodyne extension by 2.2, there is a morphism $w': L\times X\to E$ such that $u'=w'\circ(i\times X)$ and $v'=p\circ w'$. The morphism $w: L\to \mathcal{H}om(X,E)$ associated with w' is then such that $u=w\circ i$ and $v=\mathcal{H}om(X,p)\circ w$.

If we apply our proposition to the case where $B=\Delta[0]$, $\mathcal{H}om(X,B)$ is identified with $\Delta[0]$: hence, *$\mathcal{H}om(X,E)$ is a Kan complex when $\mathcal{E}$ is a Kan complex.*

3.1.3. *If $j: Y\to X$ is a monomorphism of $\Delta^\circ\mathcal{E}$ and E is a Kan complex, $\mathcal{H}om(i,E): \mathcal{H}om(X,E)\to\mathcal{H}om(Y,E)$ is a fibration.*

Consider a commutative square (Fig. 36)

$$\begin{array}{ccc} K & \xrightarrow{u} & \mathcal{H}om(X,E) \\ {\scriptstyle i}\downarrow & & \downarrow{\scriptstyle \mathcal{H}om(i,E)} \\ L & \xrightarrow{v} & \mathcal{H}om(Y,E) \end{array}$$

Fig. 36

where i is an anodyne extension. The morphisms u': $K\times X\to E$ and v': $L\times Y\to E$, canonically associated with u and v, make the following square commutative:

$$\begin{array}{ccc} K\times Y & \longrightarrow & K\times X \\ {\scriptstyle i\times Y}\downarrow & & \downarrow{\scriptstyle u'} \\ L\times Y & \longrightarrow & E \end{array}$$

and induce a morphism t: $(L\times Y)\coprod_{K\times Y}(K\times X)\to E$. If we identify K and Y with subcomplexes of L and X by means of i and j, $(L\times Y)\coprod_{K\times Y}(K\times Y)$ is simply $L\times Y\cup K\times X$. By 2.2, t can be extended to a morphism $w': L\times X\to E$; the morphism w: $L\to\mathcal{H}om(X,E)$, canonically associated with w', is then such that $u=w\circ i$ and $v=\mathcal{H}om(i,E)\circ w$.

3.1.4. *Proposition: If i: $K\to L$ is an anodyne extension and X a Kan complex, the map $\overline{\Delta^\circ\mathcal{E}}(i,X)$: $\overline{\Delta^\circ\mathcal{E}}(L,X)\to\overline{\Delta^\circ\mathcal{E}}(K,X)$ is a bijection* (with the terminology of I, 4, a Kan complex is then left closed for the set $\bar{A}$ of anodyne extensions of $\overline{\Delta^\circ\mathcal{E}}$).

It follows from the definitions that $\overline{\Delta^\circ\mathcal{E}}(i,X)$ is surjective. The proposition follows then from I, 4.1.1.

3.1.5. *Corollary: If i: $K\to L$ is an anodyne extension and X is a Kan complex, the morphism $\mathcal{H}om(i,X)$: $\mathcal{H}om(L,X)\to\mathcal{H}om(K,X)$ is a homotopy equivalence* [in other words, the image of $\mathcal{H}om(i,X)$ in $\overline{\Delta^\circ\mathcal{E}}$ is invertible].

We must show that $\overline{\Delta^\circ\mathscr{E}}(T, \mathscr{H}om(i, X))$ is a bijection for each complex T. By 1.6 $\overline{\Delta^\circ\mathscr{E}}(T, \mathscr{H}om(i, X))$ is simply $\overline{\Delta^\circ\mathscr{E}}(T\times i, X)$, which is a bijection from $\overline{\Delta^\circ\mathscr{E}}(T\times L, X)$ onto $\overline{\Delta^\circ\mathscr{E}}(T\times K, X)$ by 3.1.4 and 2.2.

3.2. Now we are able to prove the main theorem of this paragraph.

Theorem: *The canonical functor* $P_{\bar{A}}\colon \overline{\Delta^\circ\mathscr{E}} \to \mathscr{H}$ *has a right adjoint.*

By 3.1.4 and I, 4.1, it is sufficient to show that, for each complex X, there is an anodyne extension $a(X)\colon X\to X_K$ such that X_K is a Kan complex: in order to do this, let us call X-*horn* any triple $\gamma=(n, k, u)$ formed by an integer $n>1$, an integer k such that $0\leqq k\leqq n$ and a morphism $u\colon \Lambda^k[n]\to X$; we also write $n(\gamma)$, $k(\gamma)$ and $u(\gamma)$ for the components of the triple γ. We then have a diagram (Fig. 37)

$$\begin{array}{ccc} \coprod_\gamma \Lambda^{k(\gamma)}[n(\gamma)] & \xrightarrow{u(X)} & X \\ {\scriptstyle v(X)}\big\downarrow & & \\ \coprod_\gamma \Delta[n(\gamma)] & & \end{array}$$

Fig. 37

where γ runs through the X-horns, where the components of $u(X)$ are the morphisms $u(\gamma)$, and where $v(X)$ is induced by the inclusions of $\Lambda^{k(\gamma)}[n(\gamma)]$ into $\Delta[n(\gamma)]$. Since $v(X)$ is obviously an anodyne extension, the same holds for the canonical morphism $w(X)$ from X to the amalgamated sum $X_{(1)}$ of the above diagram. Let us write $X_{(2)}=X_{(1)(1)}, \ldots,$ $X_{(n+1)}=X_{(n+1)}$ and $w_n(X)=w(X_n)$; we then have an infinite sequence of anodyne extensions

$$X \xrightarrow{w(X)} X_{(1)} \xrightarrow{w_1(X)} X_{(2)} \xrightarrow{w_2(X)} X_{(3)} \to \cdots.$$

I say now that the direct limit $X_{(\infty)}$ of the $X_{(n)}$ is a Kan complex and that the canonical morphism $e(X)\colon X\to X_{(\infty)}$ is an anodyne extension: indeed, the last statement follows from 2.1 (iv); in order to prove the first one, consider a morphism $f\colon \Lambda^k[n]\to X_{(\infty)}$. Since $\Lambda^k[n]$ is a complex of finite type (II, 3.4), f factors through $X_{(p)}$, when p is large enough. It follows then from the construction of $X_{(p+1)}$ that the morphism $g\colon \Lambda^k[n]\to X_{(p)}$, induced by f, extends to a morphism $h\colon \Delta[n]\to X_{(p+1)}$; this proves the theorem.

3.2.1. Let us now associate with each complex X a Kan complex X_K and an anodyne extension $a(X)\colon X\to X_K$ (this is possible by 3.2). By I, 4, the map $f \rightsquigarrow (P_{\bar{A}}\, a(X))^{-1}(P_{\bar{A}} f)$ is a bijection

$$\overline{\Delta^\circ\mathscr{E}}(T, X_K)\to\mathscr{H}(T, K)$$

for each complex T. Moreover, if $g\colon X\to Y$ is a morphism of $\mathscr{H}$, there is one and only one morphism $g_K\colon X_K\to Y_K$ of $\overline{\Delta^\circ\mathscr{E}}$ such that $(P_{\bar{A}} a(Y))\circ g = P_{\bar{A}}(g_K\circ a(X))$. Also, the map $a\colon X\rightsquigarrow a(X)$ is an adjunction morphism from the canonical functor $P_{\bar{\Delta}}$ to the functor $X \rightsquigarrow X_K$ (I, 4.3). Finally, we can state the following theorem:

Theorem: The functor $X \rightsquigarrow X_K$ is an equivalence from $\mathscr{H}$ onto the full subcategory of $\overline{\Delta^\circ \mathscr{E}}$ formed by all Kan complexes.

3.2.2. *The canonical functor $P_{\bar{\Delta}}$: $\overline{\Delta^\circ \mathscr{E}} \to \mathscr{H}$ commutes with finite products*: let (X_i) be a finite family of complexes; by 3.1.1 (iv bis), the product $\prod_i X_{iK}$ is a Kan complex; by 2.2, the morphism $\prod_i a(X_i)$: $\prod_i X_i \to \prod_i X_{iK}$ is an anodyne extension. By 3.2.1 and 1.4, we then have

$$\mathscr{H}(T, \prod_i X_i) \simeq \overline{\Delta^\circ \mathscr{E}}(T, \prod_i X_{iK}) \simeq \prod_i \overline{\Delta^\circ \mathscr{E}}(T, X_{iK}) \simeq \prod_i \mathscr{H}(T, X_i) \qquad \text{Q.E.D.}$$

3.2.3. Let X, Y, Z be three complexes. By 1.5, we have functor isomorphisms

$$\mathscr{H}(X \times Y, Z) \simeq \overline{\Delta^\circ \mathscr{E}}(X \times Y, Z_K) \simeq \overline{\Delta^\circ \mathscr{E}}(X, \mathscr{H}om(Y, Z_K)).$$

By 3.2.1 and 3.1.2, we have

$$\overline{\Delta^\circ \mathscr{E}}(X, \mathscr{H}om(Y, Z_K)) \simeq \mathscr{H}(X, \mathscr{H}om(Y, Z_K)).$$

Moreover, using 3.1.5, we see that, passing to the quotient, the functor $(Y, Z) \rightsquigarrow \mathscr{H}om(Y, Z_K)$ from $(\overline{\Delta^\circ \mathscr{E}})^\circ \times \mathscr{H}$ to $\overline{\Delta^\circ \mathscr{E}}$ induces a functor $\mathscr{H}^\circ \times \mathscr{H}$ to $\mathscr{H}$, for which we will use the same notations. *Hence the right adjoint to the functor $X \rightsquigarrow X \times Y$ from $\mathscr{H}$ to $\mathscr{H}$ is the functor $Z \rightsquigarrow \mathscr{H}om(Y, Z_K)$.*

4. Pointed Complexes

If we want to use the results which will be proved in Chapter V, we are forced now to restate most of the statements of the beginning of this chapter in termes of pointed complexes. Since the proofs can be "copied" word for word, we will often omit them.

4.1. A pointed complex is a pair (X, x_0) formed by a simplicial set X and a 0-simplex x_0 of X; then X is called the underlying complex and x_0 the *base point*; we will often write X instead of (X, x_0): for instance, if (X, x_0) and (Y, y_0) are two pointed complexes, we will write $.\Delta^\circ \mathscr{E}(X, Y)$ for the subclass of $\Delta^\circ \mathscr{E}(X, Y)$ formed by the morphisms f such that $f(x_0) = y_0$. If (X, x_0), (Y, y_0) and (Z, z_0) are three pointed complexes and if f and g are elements of $.\Delta^\circ \mathscr{E}(X, Y)$ and $.\Delta^\circ \mathscr{E}(Y, Z)$ respectively, it is clear that the composition $g \circ f$ belongs to $.\Delta^\circ \mathscr{E}(X, Z)$; this composition law enables us to define *the category $.\Delta^\circ \mathscr{E}$ of pointed complexes:* its objects are the pointed complexes, and its set of morphisms are the sets $.\Delta^\circ \mathscr{E}(X, Y)$ (see II, 7.4).

When we will speak about the pointed complex $\Delta[1]$ it will be tacitly assumed that the base point is the map from $[0]$ to $[1]$ which sends 0 to 0; similarly, we will still write $\dot{\Delta}[1]$ for the pointed complex whose underlying complex is the boundary of $\Delta[1]$, and which has the same base point as $\Delta[1]$. Finally, we will make the *circle* Ω a pointed complex by choosing the unique vertex of this complex as its base point (see II, 2.5.2).

4.1.1. It is clear that each small diagram of $.\Delta^\circ\mathscr{E}$ has a direct and an inverse limit. Let us say simply that the *direct* product of two pointed complexes (X, x_0) and (Y, y_0) has the product $X \times Y$ as its underlying complex, and the pair (x_0, y_0) as its base point. Similarly the complex underlying to the direct sum $X \vee Y$ of (X, x_0) and (Y, y_0) is the amalgamated sum of the diagram

$$\begin{array}{ccc} \Delta[0] & \xrightarrow{\tilde{y}_0} & Y \\ {\scriptstyle \tilde{x}_0}\downarrow & & \\ X & & \end{array}$$

of $\Delta^\circ\mathscr{E}$, and its base point is the common image of x_0 and y_0.

$\Delta[0]$ is obviously the zero object of the category $.\Delta^\circ\mathscr{E}$; if (X, x_0) and (Y, y_0) are two pointed complexes, there is one and only one morphism from (X, x_0) to (Y, y_0) which factors through $\Delta[0]$; it will be written 0^X_Y, and it will be called the *zero morphism*. Then the morphism from X to $X \times Y$ whose components are $\operatorname{Id} X$ and $0^X_{X\times Y}$, and the morphism from Y to $X \times Y$ whose components are $0^Y_{X\times Y}$ and $\operatorname{Id} Y$, induce a monomorphism

$$i_{X,Y}\colon X \vee Y \to X \times Y.$$

The image of $i_{X,Y}$ is the subcomplex $X \times \{y_0\} \cup \{x_0\} \times Y$ of $X \times Y$[1].

In the sequel, we will write $X \wedge Y$ for the pointed complex whose underlying complex is the amalgamated sum of the diagram

$$\begin{array}{ccc} X \vee Y & \xrightarrow{i_{X,Y}} & X \times Y \\ \downarrow & & \\ \Delta[0] & & \end{array}$$

of $\Delta^\circ\mathscr{E}$, and whose base point is the image of (x_0, y_0). It is easily seen that the associativity and commutativity properties of the direct product are transposed to the *contracted product* $X \wedge Y$.

4.1.2. Let (Y, y_0) and (Z, z_0) be two pointed complexes. We will write $\mathscr{H}\!om.(Y, Z)$ for the subcomplex of $\mathscr{H}\!om(Y, Z)$ whose n-simplices are the morphisms $f\colon \Delta[n] \times Y \to Z$ which send $\Delta[n] \times \{y_0\}$ to $\{z_0\}$; we make $\mathscr{H}\!om.(Y, Z)$ a pointed complex by taking the zero morphism 0^Y_Z as the base point. For example, if Y is equal to $\dot{\Delta}[1]$, a morphism

[1] $\{x_0\}$ denotes the image of $\Delta[0]$ under the morphism $\tilde{x}_0\colon \Delta[0] \to X$.

$f\colon \Delta[n]\times\dot{\Delta}[1]\to Z$ which sends $\Delta[n]\times\{0\}$ to $\{z_0\}$ is characterized by its restriction to $\Delta[n]\times\{1\}\simeq\Delta[n]$; hence we can identify $\mathcal{H}om.(\dot{\Delta}[1],Z)$ with Z.

Consider now three pointed complexes (X, x_0), (Y, y_0) and (Z, z_0). The set $.\Delta^\circ\mathcal{E}(X, \mathcal{H}om.(Y,Z))$ is then the subset of $\Delta^\circ\mathcal{E}(X, \mathcal{H}om(Y,Z))$ formed by all morphisms x such that the diagram of Fig. 38

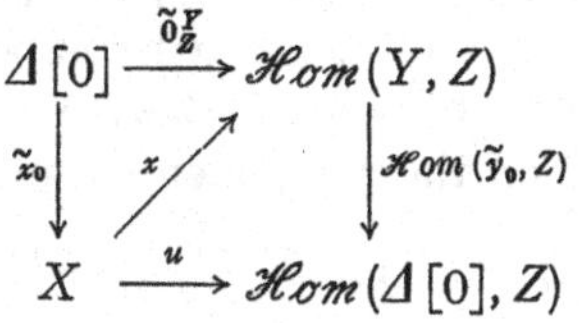

Fig. 38

of $\Delta^\circ\mathcal{E}$ is commutative (u is the zero morphism when we provide $\mathcal{H}om(\Delta[0],Z)$ with the base point $\tilde{z}_0$). The isomorphism φ^{-1} of II, 2.5.3 associates with x a morphism x' which makes the diagram of Fig. 39 commutative:

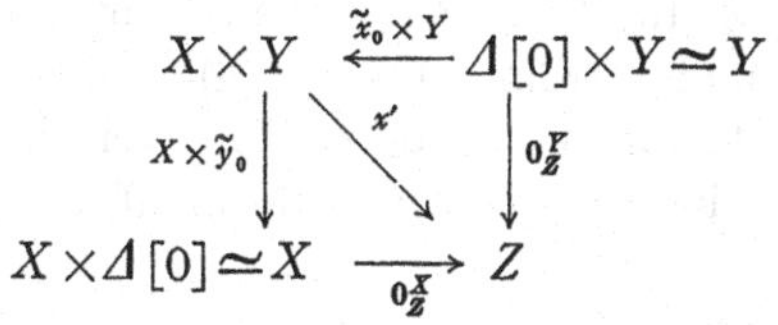

Fig. 39

Hence the morphism x' induces the zero morphism on $X\times\{y_0\}\cup\{x_0\}\times Y$; in other words, x' factors through $X\wedge Y$. It follows that the functor isomorphism φ induces an isomorphism

$$(*)\qquad .\varphi\colon .\Delta^\circ\mathcal{E}(X\wedge Y, Z) \xrightarrow{\sim} .\Delta^\circ\mathcal{E}\big(X, \mathcal{H}om.(Y,Z)\big).$$

Now, if we consider the complex $\Delta'[n]=\Delta[n]\amalg\Delta[0]$ pointed by $\Delta[0]$, and if A is a pointed complex we have the formula: $.\Delta^\circ\mathcal{E}(\Delta'[n],A)\simeq A_n$, so that, according to $(*)$ the following isomorphism holds:

$$(**)\qquad .\Delta^\circ\mathcal{E}(\Delta'[n]\wedge Y, Z)\simeq\mathcal{H}om.(Y,Z)_n.$$

Finally we get isomorphisms $\mathcal{H}om.(X\wedge Y,Z)_n\simeq .\Delta^\circ\mathcal{E}(\Delta'[n]\wedge X\wedge Y,Z)\simeq .\Delta^\circ\mathcal{E}\big(\Delta'[n]\wedge X, \mathcal{H}om.(Y,Z)\big)\simeq\mathcal{H}om.\big(X,\mathcal{H}om.(Y,Z)\big)_n$ where the first and the third are given by $(**)$ and the second by $(*)$.

Thus the isomorphism $.\varphi$ induces an isomorphism

$$.\Phi\colon \mathcal{H}om.(X\wedge Y,Z)\xrightarrow{\sim}\mathcal{H}om.\big(X,\mathcal{H}om.(Y,Z)\big).$$

4.2. The vertices of $\mathcal{H}om.(Y,Z)$ are identified with the morphisms from (Y, y_0) to (Z, z_0). If f and g are two such morphisms, a 1-simplex h

of $\mathcal{H}om.(Y,Z)$ such that $d_1 h = f$ and $d_0 h = g$ is a homotopy h connecting f with g which sends $\Delta[1]\times\{y_0\}$ to $\{z_0\}$: we will say that h *respects base points*. Two morphisms f and g from (Y, y_0) to (Z, z_0) are called *homotopic relatively to base points* or simply homotopic if they belong to the same connected component of $\mathcal{H}om.(Y,Z)$; it is equivalent to say that there is a composed homotopy $h: I_n\times Y\to Z$ connecting f with g which respects base points, i.e. such that h sends $I_n\times\{y_0\}$ to $\{z_0\}$.

Let $\Pi_0^{\cdot}(Y,Z)$ be the set of connected components of $\mathcal{H}om.(Y,Z)$. The maps $\nu_{X,Y,Z}$ of 1.3 sends $\mathcal{H}om.(X,Y)\times\mathcal{H}om.(Y,Z)$ into $\mathcal{H}om.(X,Z)$ and induce on the components a map

$$\mu^{\cdot}_{X,Y,Z}: \Pi_0^{\cdot}(X,Y)\times\Pi_0^{\cdot}(Y,Z)\to\Pi_0^{\cdot}(X,Z).$$

We will call category *of pointed complexes modulo homotopy* (noted $.\overline{\Delta^\circ\mathcal{E}}$) the category whose objects are the pointed complexes, whose sets of morphisms are the $\Pi_0^{\cdot}(Y,Z)$, and whose composition laws are the maps $\mu^{\cdot}_{X,Y,Z}$; we will call *canonical functor* from $.\Delta^\circ\mathcal{E}$ to $.\overline{\Delta^\circ\mathcal{E}}$ the functor which induces the identity on the set of objects and the canonical maps from $.\Delta^\circ\mathcal{E}(Y,Z)$ to $\Pi_0^{\cdot}(Y,Z)$ on the sets of morphisms.

The canonical functor from $.\Delta^\circ\mathcal{E}$ to $.\overline{\Delta^\circ\mathcal{E}}$ commutes with finite direct sums and finite direct products. Moreover, passing to the connected components, the functor isomorphism $.\Phi$ of 4.1.2 gives an isomorphism

$$\Pi_0^{\cdot}(\Phi): .\overline{\Delta^\circ\mathcal{E}}(X\wedge Y,Z)\simeq .\overline{\Delta^\circ\mathcal{E}}(X,\mathcal{H}om.(Y,Z)).$$

If $f, g: X\rightrightarrows X'$ are two homotopic morphisms of $.\Delta^\circ\mathcal{E}$, we obviously have $.\overline{\Delta^\circ\mathcal{E}}(f,\mathcal{H}om.(Y,Z)) = .\overline{\Delta^\circ\mathcal{E}}(g,\mathcal{H}om.(Y,Z))$, hence

$$.\overline{\Delta^\circ\mathcal{E}}(f\wedge Y,Z) = .\overline{\Delta^\circ\mathcal{E}}(g\wedge Y,Z)$$

for all Y and Z; hence $f\wedge Y$ and $g\wedge Y$ are homotopic; since the contracted product is "symmetric", $X\wedge e$ and $X\wedge d$ are also homotopic if e and d are homotopic. Hence the functor $(X,Y)\rightsquigarrow X\wedge Y$ from $.\Delta^\circ\mathcal{E}\times.\Delta^\circ\mathcal{E}$ to $.\Delta^\circ\mathcal{E}$ induces a functor from $.\overline{\Delta^\circ\mathcal{E}}\times.\overline{\Delta^\circ\mathcal{E}}$ to $.\overline{\Delta^\circ\mathcal{E}}$ which will be denote by the same symbol $\wedge$. Similarly, the functor $\mathcal{H}om.$ from $(.\Delta^\circ\mathcal{E})^\circ\times(.\Delta^\circ\mathcal{E})$ to $.\Delta^\circ\mathcal{E}$ induces a functor, still written $\mathcal{H}om.$, from $(.\overline{\Delta^\circ\mathcal{E}})^\circ\times(.\overline{\Delta^\circ\mathcal{E}})$ to $.\overline{\Delta^\circ\mathcal{E}}$.

4.3. A morphism f of $.\overline{\Delta^\circ\mathcal{E}}$ is *an anodyne extension* if the image of f under the canonical functor $(X,x_0)\rightsquigarrow X$ from $.\overline{\Delta^\circ\mathcal{E}}$ to $.\overline{\Delta^\circ\mathcal{E}}$ is an anodyne extension (2.3). *The set $.\bar{A}$ of anodyne extensions of $.\overline{\Delta^\circ\mathcal{E}}$ admits a calculus of left fractions;* the corresponding category of fractions will be called the *pointed homotopic category*, and will be written $.\mathcal{H}$; if (X,x_0) and (Y,y_0) are two pointed complexes, we will write $.\mathcal{H}(X,Y)$ instead of $.\mathcal{H}((X,x_0),(Y,y_0))$.

The canonical functor from $.\overline{\Delta^\circ\mathscr{E}}$ to $.\mathscr{H}$ obviously commutes with finite direct sums, for which we will still use the symbol $\vee$. Similarly, *for each pointed complex Z and each anodyne extension i: $Y \to X$ of $.\overline{\Delta^\circ\mathscr{E}}$, the morphism $Z \wedge i$: $Z \wedge Y \to Z \wedge X$* is an anodyne extension: for let j: $Y \to X$ be an element of $.\Delta^\circ\mathscr{E}(Y, X)$ whose image in $.\overline{\Delta^\circ\mathscr{E}}(Y, X)$ is i. We can obviously suppose that j is the inclusion into X of a subcomplex Y; in that case, we have a commutative diagram (Fig. 40)

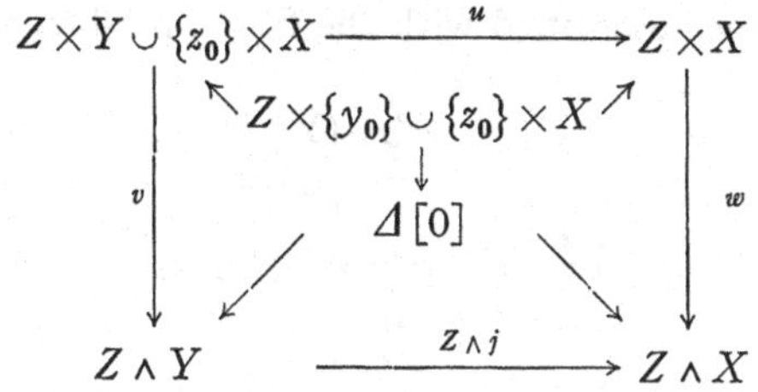

Fig. 40

of $\Delta^\circ\mathscr{E}$, where u is the inclusion, w the canonical projection, and where v induces the canonical projection on $Z \times Y$, and the zero-morphism on $\{z_0\} \times X$. The two interior squares are cocartesian by definition of the contracted product; hence the same holds for the large square. Since u is an anodyne extension by 2.2, the same holds for $Z \wedge j$ [2.1 (ii)].

In a similar way, we see that $i \wedge Z$ is an anodyne extension if i is one. Consequently, by passing to the categories of fractions, the functor $(X, Y) \rightsquigarrow X \wedge Y$ from $(.\overline{\Delta^\circ\mathscr{E}}) \times (.\overline{\Delta^\circ\mathscr{E}})$ to $.\overline{\Delta^\circ\mathscr{E}}$ defines a functor from $.\mathscr{H} \times .\mathscr{H}$ to $.\mathscr{H}$, for which we will obviously use the same symbol $\wedge$.

4.3.1. Let us now associate with each pointed complex X a pointed Kan complex X_K and an anodyne extension $a(X)$: $X \to X_K$ of $.\overline{\Delta^\circ\mathscr{E}}$ (this is possible by 3.2). If p is the canonical functor from $.\overline{\Delta^\circ\mathscr{E}}$ to the category of fractions $.\mathscr{H}$, the map $f \rightsquigarrow (p\, a(X))^{-1} \circ (p\, f)$ is a bijection

$$.\overline{\Delta^\circ\mathscr{E}}(T, X_K) \simeq .\mathscr{H}(T, X)$$

for each pointed complex T. Moreover, if g: $X \to Y$ is a morphism of $.\mathscr{H}$, there is one and only one morphism g_K: $X_K \to Y_K$ of $.\overline{\Delta^\circ\mathscr{E}}$ such that $(p\, a(Y)) \circ g = p\,(g_K \circ a(X))$. Thus we define *a functor $X \rightsquigarrow X_K$ which is right adjoint to the canonical functor p: $.\overline{\Delta^\circ\mathscr{E}} \to .\mathscr{H}$ and which induces an equivalence from the pointed homotopic category $.\mathscr{H}$ onto the full subcategory of $.\overline{\Delta^\circ\mathscr{E}}$ formed by all pointed Kan complexes.*

Finally we see, as in 3.2.2 and 3.2.3 that the canonical functor p: $.\Delta^\circ\mathscr{E} \to .\mathscr{H}$ commutes with finite products. Moreover, passing to the quotient, the functor $(Y, Z) \rightsquigarrow \mathscr{H}om.(Y, Z_K)$ from $(.\overline{\Delta^\circ\mathscr{E}})^\circ \times .\mathscr{H}$ to $.\overline{\Delta^\circ\mathscr{E}}$ induces a functor from $.\mathscr{H}^\circ \times .\mathscr{H}$ to $.\mathscr{H}$, for which we will use the same notations. *Hence the functor $Z \rightsquigarrow \mathscr{H}om.(Y, Z_K)$ is right adjoint to* *the functor* $X \rightsquigarrow X \wedge Y$ *from* $.\mathscr{H}$ *to* $.\mathscr{H}$. This last statement follows from the "equalities"

$$.\mathscr{H}(X \wedge Y, Z) \simeq .\overline{\Delta^\circ\mathscr{E}}(X \wedge Y, Z_K) \simeq .\overline{\Delta^\circ\mathscr{E}}(X, \mathscr{H}om.(Y, Z_K)) \simeq .\mathscr{H}(X, \mathscr{H}om.(Y, Z_K)).$$

The last equality follows from the fact that $\mathscr{H}om.(Y, T)$ *is a Kan complex if* (T, t_0) *is a pointed Kan complex. By the* definition of 4.1.2, $\mathscr{H}om.(Y, T)$ is indeed identified with the fibred product of the diagram of Fig. 41

$$\begin{array}{ccc} & & \mathscr{H}om(Y, T) \\ & & \big\downarrow{\scriptstyle \mathscr{H}om(\tilde{y}_0, T)} \\ \Delta[0] & \xrightarrow{\tilde{t}_0} & T \simeq \mathscr{H}om(\Delta[0], T) \end{array}$$

Fig. 41

so that our statement follows from 3.1.3 and 3.1.1 (ii bis).

4.4. *Remark*: Let (X, x_0) be a pointed complex; for X_K, take the complex $X_{(\infty)}$ constructed in 3.2, with base point $e(X)(x_0)$. Then $X \rightsquigarrow X_K$ is a functor from $.\Delta^\circ\mathscr{E}$ to $.\Delta^\circ\mathscr{E}$ and e is a functor morphism from $\mathrm{Id}(.\Delta^\circ\mathscr{E})$ to $?_K$: if $f\colon X \to Y$ is a morphism of $.\Delta^\circ\mathscr{E}$, we have the following commutative diagram of $.\Delta^\circ\mathscr{E}$:

$$\begin{array}{ccc} X & \xrightarrow{e(X)} & X_K \\ {\scriptstyle f}\big\downarrow & & \big\downarrow{\scriptstyle f_K} \\ Y & \xrightarrow{e(Y)} & Y_K \end{array}$$

In the sequel, we will say that $(e, ?_K)$ is a *Kan envelope*.

4.5. As a premium, and also as an exercise in the calculus of fractions, we will now show that *we can give to the simplicial circle* Ω *a co-group structure in the pointed homotopic category* $.\mathscr{H}$: we know that such a structure is given by a morphism

$$\varphi\colon \Omega \to \Omega \vee \Omega$$

such that, for each pointed complex T, the induced map $.\mathscr{H}(\varphi, T)$: $.\mathscr{H}(\Omega \vee \Omega, T) \simeq .\mathscr{H}(\Omega, T) \times .\mathscr{H}(\Omega, T) \to .\mathscr{H}(\Omega, T)$ is the composition law of a group structure on the set $.\mathscr{H}(\Omega, T)$. It is equivalent to say that φ satisfies the following three conditions:

(i) The diagram

$$\begin{array}{ccc} \Omega & \xrightarrow{\varphi} & \Omega \vee \Omega \\ {\scriptstyle \varphi}\big\downarrow & & \big\downarrow{\scriptstyle \Omega \vee \varphi} \\ \Omega \vee \Omega & \xrightarrow{\varphi \vee \Omega} & \Omega \vee \Omega \vee \Omega \end{array}$$

is commutative.

(ii) If $\psi\colon \Omega \vee \Omega \to \Omega$ is the morphism whose components are the zero morphism and the identity morphism, then the composition

$$\Omega \xrightarrow{\varphi} \Omega \vee \Omega \xrightarrow{\psi} \Omega$$

is the identity morphism of Ω.

(iii) There exists a morphism $c\colon \Omega \to \Omega$ such that the composition

$$\Omega \xrightarrow{\varphi} \Omega \vee \Omega \xrightarrow{c \vee \Omega} \Omega \vee \Omega \xrightarrow{\delta} \Omega$$

is the zero morphism (here δ is the codiagonal morphism whose components are both equal to the identity morphism of Ω).

In order to define φ, let us write $\widetilde{X}$ for the amalgamated sum of a diagram of the form

$$\begin{array}{ccc} Sk^0 X & \xrightarrow{\text{inclusion}} & X \\ \downarrow & & \\ \Delta[0] & & \end{array}$$

where X is an arbitrary simplicial set; we have for example, $\Omega = \widetilde{\Delta[1]}$. By passing to the quotient, any morphism $f\colon X \to Y$ obviously induces a morphism $\tilde{f}\colon \widetilde{X} \to \widetilde{Y}$; if we consider $\widetilde{X}$ and $\widetilde{Y}$ as pointed complexes (they have only one vertex), $\tilde{f}$ is even a morphism of pointed complexes. Consider then the following diagram of $.\Delta^\circ\mathscr{E}$ (Fig. 42)

$$\begin{array}{ccc} \Omega \simeq \widetilde{\Delta[1]} & & \Omega \vee \Omega \\ & \searrow^{\widetilde{\Delta(\partial^1_2)}} & \downarrow s \\ & & \widetilde{\Delta[2]} \end{array}$$

Fig. 42

where s is the morphism whose components are $\widetilde{\Delta(\partial^2_2)}$ and $\widetilde{\Delta(\partial^2_0)}$. The pointed direct sum $\Omega \vee \Omega$ is obviously identified with $\widetilde{\Lambda^1[2]}$, and s with the indusion of $\widetilde{\Lambda^1[2]}$ into $\widetilde{\Delta[2]}$. Since the square of Fig. 43

$$\begin{array}{ccc} \Lambda^1[2] & \xrightarrow{\text{can.}} & \widetilde{\Lambda^1[2]} \\ {\scriptstyle \text{incl.}}\downarrow & & \downarrow{\scriptstyle \text{incl.}} \\ \Delta[2] & \xrightarrow{\text{can.}} & \widetilde{\Delta[2]} \end{array}$$

Fig. 43

is cocartesian, we see that s is an anodyne extension. Hence, if q is the canonical functor from $.\overline{\Delta^\circ\mathscr{E}}$ to $.\mathscr{H}$, we can define $\varphi = (q\, s)^{-1} \circ q\big(\widetilde{\Delta(\partial^1_2)}\big)$ (see the description of the morphisms of a category of fractions, given in I, 2.3).

It remains to be seen that φ satisfies conditions (i), (ii) and (iii) given above. Since this can also be deduced from further considerations, the verification is left to the reader (see 5.5).

5. Poincaré Group of a Pointed Complex

Let (X, x_0) be a pointed complex. We will show that *the Poincaré group* $\Pi_1(X, x_0)$ *defined in* II, 7.4. *is simply the set* $\mathcal{H}(\Omega, X)$ *of morphisms from the simplicial circle to* X.

5.1. Let us show first that, if X is a Kan complex, the image of X_1 in the set of morphisms of the groupoid $\Pi(X)$ is stable under inversion and composition; in other words, this image is the set of all morphisms.

Indeed, let α be a morphism of $\Pi(X)$. By II, 7.2, α is the image of a morphism $f\colon I_n \to X$. On the other hand, let $i\colon I_n \to \Delta[n]$ be the morphism associated with the following increasing map (Fig. 44)

$$\begin{array}{ccccccccc} 0 & < & 1 & > & 2 & < & 3 & > & 4 \ldots \\ \downarrow & & \downarrow & & \downarrow & & \downarrow & & \\ 0 & < & n & > & 1 & < & n-1 & > & 2 \ldots \end{array}$$

Fig. 44

This morphism i is obviously an anodyne extension, so that the diagram

$$\begin{array}{ccc} I_n & \xrightarrow{f} & X \\ {\scriptstyle i}\downarrow & & \\ \Delta[n] & & \end{array}$$

can be completed by a morphism $F\colon \Delta[n] \to X$ such that $F \circ i = f$. If $\Delta(j)\colon \Delta[1] \to \Delta[n]$ is the morphism associated with the increasing map $j\colon [1] \to [n]$ which sends 0 to 0 and 1 to the integral part of $\frac{n+1}{2}$, the image of the edge of X defined by $F \circ \Delta(j)$ in the set of morphisms of $\Pi(X)$ is simply α.

5.2. *The Poincaré groupoid of a Kan complex* X *can be described as follows:*

The set of objects is the set X_0 *of* 0*-simplices of* X.

The set of morphisms is the quotient of X_1 *by an equivalence relation such that two* 1*-simplices* s *and* t *are equivalent if and only if the following conditions are satisfied*: $d_0 s = d_0 t$; $d_1 s = d_1 t$; *there is a homotopy* h: $\Delta[1] \times \Delta[1] \to X$ *connecting the singular simplices* $\tilde{s}$ *and* $\tilde{t}$, *whose restriction to* $\Delta[1] \times \dot{\Delta}[1]$ *factors through the canonical projection of* $\Delta[1] \times \dot{\Delta}[1]$ *onto* $\dot{\Delta}[1]$.

The domain and range maps are induced by d_1 *and* d_0.

If $[s]$ *is the equivalence class of* s *under the above relation, the composition of morphisms is such that* $[d_1 \sigma] = [d_0 \sigma] \circ [d_2 \sigma]$ *for each* 2*-simplex* σ *of* X.

It is clear that the above conditions define a groupoid, which will be temporarily noted $G(X)$. We see then that $G(X)$ is identified with $\Pi(X)$: for let $\Gamma(X)$ be the groupoid of paths of the diagram scheme $d_1, d_0\colon X_1 \rightrightarrows X_0$. The identity map of X_0 and the canonical map from X_1 to the set of morphisms of $G(X)$ induce a functor $p\colon \Gamma(X) \to G(X)$. Since p is obviously compatible with the relations which defined the quotient $\Pi(X)$ of $\Gamma(X)$ (II, 7.1), p induces a functor $q\colon \Pi(X) \to G(X)$.

This functor induces a surjection on morphisms. Let then α and β be two morphisms of $\Pi(X)$ such that $q\,\alpha = q\,\beta$. By 5.1, α and β are the images of two 1-simplices s and t of X; the equality $q\,\alpha = q\,\beta$ is then equivalent to $[s] = [t]$, i.e. to the existence of a $\sigma \in X_2$ such that $d_2\,\sigma = s$, $d_1\,\sigma = t$, $d_0\,\sigma = s_0\,d_0\,s = s_0\,d_0\,t$. By II, 7.1, we then have $\alpha = \beta$, so that q is an isomorphism.

5.3. The description given in 5.2 can be applied in particular to the complex $\mathscr{H}\!om(T, X)$ when X is a Kan complex (3.1.2). In that case, *two homotopic morphisms* f, g: $T \to X$, i.e. *which are connected by a composed homotopy, are already connected by a homotopy* $h\colon \Delta[1] \times T \to X$ (1.1).

Similarly, if (T, t_0) and (X, x_0) are pointed complexes, and if X is a Kan complex, $\mathscr{H}\!om.(T, X)$ is a Kan complex (4.3.1). *Hence two morphisms* $f, g\colon T \rightrightarrows X$ *are homotopic relatively to* t_0 *and* x_0 *if there is a homotopy* $h\colon \Delta[1] \times T \to X$ *connecting* f *with* g, *which sends* $\Delta[1] \times \{t_0\}$ *to* $\{x_0\}$.

5.4. Let us now use 5.2 for the description of the Poincaré group of a pointed Kan complex (X, x_0): let X_1^1 be the set of 1-simplices s of X such that $d_0\,s = d_1\,s = x_0$. The singular simplex $\tilde{s}\colon \Delta[1] \to X$ associated with such an s factors through the canonical projection $p\colon \Delta[1] \to \Omega$, so that $\tilde{s}$ is of the form $s' \circ p$.

It is clear that the map $s \rightsquigarrow s'$ is a bijection from X_1^1 onto $.\Delta^\circ\mathscr{E}(\Omega, X)$. This bijection is compatible with the equivalence relations defined in 5.2 and 4.2. Hence it induces a bijection of the quotients $\Pi_1(X, x_0)$ and $.\overline{\Delta^\circ\mathscr{E}}(\Omega, X)$. Moreover, since X is an Kan complex, the canonical map from $.\overline{\Delta^\circ\mathscr{E}}(\Omega, X)$ to $.\mathscr{H}(\Omega, X)$ is a bijection. Thus we obtain the isomorphism from $.\mathscr{H}(\Omega, X)$ onto $\Pi_1(X, x_0)$ we were looking for.

5.5. Finally, consider an arbitrary pointed complex (X, x_0): we then have the diagram of Fig. 45, where $(e, ?_K)$ is a Kan envelope (4.4) and where the vertical arrow is the isomorphism described in 5.4:

$$\begin{array}{ccc}
.\mathscr{H}(\Omega, X) & \xrightarrow{.\mathscr{H}(\Omega,\, e(X))} & .\mathscr{H}(\Omega, X_K) \\
 & & \downarrow \wr \\
\Pi_1(X, x_0) & \xrightarrow{\Pi_1(e(X))} & \Pi_1\big(X_K, e(X)(x_0)\big)
\end{array}$$

Fig. 45

The map $.\mathscr{H}(\Omega, e(X))$ is obviously bijective [$e(X)$ is an anodyne extension]. On the other hand, the morphism $v(X)$ of 3.2 induces an isomorphism between the Poincaré groupoids. Since the functor Π commutes with direct limits, it follows that $w(X)$ and $e(X)$ induce isomorphisms from $\Pi(X)$ onto $\Pi(X_{(1)})$ and $\Pi(X_K)$. In particular, $\Pi_1(e(X))$ is a bijection.

Thus we obtain by composition an isomorphism from $.\mathscr{H}(\Omega, X)$ onto $\Pi_1(X, x_0)$. This isomorphism is functorial in X, and gives to the sets $.\mathscr{H}(\Omega, X)$ group structures which are functorial in X. Hence Ω has a cogroup structure. It follows from the definitions that the comultiplication law of Ω is the one defined in 4.5.

Chapter Five

Exact Sequences of Algebraic Topology

We give here a unified account of some exact sequences which we meet in algebraic topology. The proofs we give do not modify in an essential way those of PUPPE. They are more abstract, selfdual and sometimes more simple (?).

We write $\mathbb{0}$ for the *zero category* whose set of objects and whose set of morphisms both are the cardinal number 1. For each object x of a category $\mathscr{X}$ we identify x, as usual, with the functor from $\mathbb{0}$ to $\mathscr{X}$ which sends the object and the morphism of $\mathbb{0}$ to x and Id x respectively.

1. 2-Categories

1.1. A 2-category $\mathfrak{C}$ is defined by the following:

a set $\mathfrak{C}_0$, called the set of objects of $\mathfrak{C}$.

a family of categories $\mathscr{H}om_{\mathfrak{C}}(x, y)$, indexed by $\mathfrak{C}_0 \times \mathfrak{C}_0$.

an object of $\mathscr{H}om_{\mathfrak{C}}(x, y)$ is called a 1-morphism of $\mathfrak{C}$;

a morphism of $\mathscr{H}om_{\mathfrak{C}}(x, y)$ is a 2-morphism of $\mathfrak{C}$.

a family of functors

$$\mu_{x,y,z}\colon \mathscr{H}om_{\mathfrak{C}}(x, y) \times \mathscr{H}om_{\mathfrak{C}}(y, z) \to \mathscr{H}om_{\mathfrak{C}}(x, z)$$

indexed by $\mathfrak{C}_0 \times \mathfrak{C}_0 \times \mathfrak{C}_0$ and called *composition functors*. The above conditions are subject to the following axioms.

for each 4-tuple (x, y, z, t) of objects of $\mathfrak{C}$, the square (Fig. 46)

$$\begin{array}{ccc} \mathscr{H}om_{\mathfrak{C}}(x, y) \times \mathscr{H}om_{\mathfrak{C}}(y, z) \times \mathscr{H}om_{\mathfrak{C}}(z, t) & \xrightarrow{\mathrm{Id} \times \mu_{y,z,t}} & \mathscr{H}om_{\mathfrak{C}}(x, y) \times \mathscr{H}om_{\mathfrak{C}}(y, t) \\ \Big\downarrow{\scriptstyle \mu_{x,y,z} \times \mathrm{Id}} & & \Big\downarrow{\scriptstyle \mu_{x,y,t}} \\ \mathscr{H}om_{\mathfrak{C}}(x, z) \times \mathscr{H}om_{\mathfrak{C}}(z, t) & \xrightarrow{\mu_{x,z,t}} & \mathscr{H}om_{\mathfrak{C}}(x, t) \end{array}$$

Fig. 46

is commutative (associativity axiom).

for each object x of $\mathfrak{C}$, there is an object i_x of $\mathcal{H}om_{\mathfrak{C}}(x, y)$ such that the triangles of Fig. 47.

$$\begin{array}{ccc} \mathcal{O} \times \mathcal{H}om_{\mathfrak{C}}(x, y) & \xrightarrow{i_x \times \mathrm{Id}} & \mathcal{H}om_{\mathfrak{C}}(x, x) \times \mathcal{H}om_{\mathfrak{C}}(x, y) \\ & \searrow^{\sim} \quad \swarrow_{\mu_{x,x,y}} & \\ & \mathcal{H}om_{\mathfrak{C}}(x, y) & \end{array}$$

and

$$\begin{array}{ccc} \mathcal{H}om_{\mathfrak{C}}(y, x) \times \mathcal{O} & \xrightarrow{\mathrm{Id} \times i_x} & \mathcal{H}om_{\mathfrak{C}}(y, x) \times \mathcal{H}om_{\mathfrak{C}}(x, x) \\ & \searrow^{\sim} \quad \swarrow_{\mu_{y,x,x}} & \\ & \mathcal{H}om_{\mathfrak{C}}(y, x) & \end{array}$$

Fig. 47

are commutative for each object y of $\mathfrak{C}$ (neutral objects axiom).

If z and t are two objects of $\mathfrak{C}$, we will often write z^t instead of $\mathcal{H}om_{\mathfrak{C}}(t, z)$ for convenience; if $f: x \to y$ is a 1-morphism, f^t will then be the functor

$$\mathcal{H}om_{\mathfrak{C}}(t, f): \mathcal{H}om_{\mathfrak{C}}(t, x) \to \mathcal{H}om_{\mathfrak{C}}(t, y)$$

induced by the composition functor $\mu_{t,x,y}$; if $\alpha: f \to f'$ is a 2-morphism we will also write α^t instead of

$$\mathcal{H}om_{\mathfrak{C}}(t, \alpha): \mathcal{H}om_{\mathfrak{C}}(t, f) \to \mathcal{H}om_{\mathfrak{C}}(t, f').$$

Thus, for each object a of the category $\mathcal{H}om_{\mathfrak{C}}(t, x)$, α^t is the functor morphism $\mu_{t,x,y}(\mathrm{Id}\, a, \alpha)$.

1.2. The "best-known" 2-category is the category of categories, which we denote by $\mathfrak{Cat}$: the objects of $\mathfrak{Cat}$ are the categories belonging to the universe $\mathcal{U}$, fixed once for all; if x and y are two such categories, $\mathcal{H}om_{\mathfrak{Cat}}(x, y)$ is the category of functors from x to y, and the composition functors are the usual functors. There are obviously many categories akin to this one: instead of all categories, we can take additive categories only, instead of all functors, additive functors only ....

1.3. If $\mathfrak{C}$ is a 2-category, we write $\mathfrak{C}_{1,2}$ for the direct sum of the categories $\mathcal{H}om_{\mathfrak{C}}(x, y)$, $\mathfrak{C}_1$ for the set of objects of $\mathfrak{C}_{1,2}$ (the 1-morphisms of $\mathfrak{C}$), $\mathfrak{C}_2$ for the set of morphisms of $\mathfrak{C}_{1,2}$ (the 2-morphisms of $\mathfrak{C}$), $\mathfrak{d}_1$ and $\mathfrak{r}_1$ for the domain and range maps of $\mathfrak{C}_{1,2}$ and $\mathfrak{i}_1$ for the map from $\mathfrak{C}_1$ to $\mathfrak{C}_2$ which associates with each object of $\mathfrak{C}_{1,2}$ the identity morphism of this object. If α and β are two composable morphisms of $\mathfrak{C}_{1,2}(\mathfrak{d}_1 \alpha = \mathfrak{r}_1 \beta)$, we write $\mu_{1,2}(\beta, \alpha)$ or $\alpha \circ \beta$ for their product.

Similarly, we write $\mathfrak{C}_{0,1}$ for the category whose set of objects is $\mathfrak{C}_0$, and such that the set of morphisms from an object x to an object y is the set of objects of $\mathcal{H}om_{\mathfrak{C}}(x, y)$, composition maps beeing induced by the functors $\mu_{x,y,z}$. We write $\mathfrak{d}_0$ and $\mathfrak{r}_0$ for the domain and range maps of $\mathfrak{C}_{0,1}$, and $\mathfrak{i}_0$ for the map $x \rightsquigarrow i_x$ defined above. It f and g are two composable morphisms of $\mathfrak{C}_{0,1}$ ($\mathfrak{d}_0 g = \mathfrak{r}_0 f$), we write $\mu_{0,1}(f, g)$, $g \cdot f$ or $g f$ for their product.

Finally, we write $\mathfrak{C}_{0,2}$ for the following category: the set of its object is $\mathfrak{C}_0$, the set of its morphisms is $\mathfrak{C}_2$, its range and domain maps are $\mathfrak{d}_0 \mathfrak{d}_1 = \mathfrak{d}_0 \mathfrak{r}_1$ and $\mathfrak{r}_0 \mathfrak{d}_1 = \mathfrak{r}_0 \mathfrak{r}_1$, and its composition maps are induced by the functors $\mu_{x,y,z}$. If α and α' are two composable maps of $\mathfrak{C}_{0,2}$, we write $\mu_{0,2}(\alpha, \alpha')$, $\alpha' * \alpha$ or $\alpha' \alpha$ for their product. In the case where α' (resp. α) is of the form $\mathfrak{i}_1(f)$, we will often write $f\alpha$ or $f * \alpha$ (resp. $\alpha' f$ or $\alpha' * f$) instead of $\alpha' * \alpha$.

These different categories are related by the following conditions:

a) If α, α', β, β' are four 2-morphisms such that

$$\mathfrak{r}_1 \alpha = \mathfrak{d}_1 \beta, \; \mathfrak{r}_1 \alpha' = \mathfrak{d}_1 \beta'$$

and

$$\mathfrak{r}_0 \mathfrak{r}_1 \alpha = \mathfrak{r}_0 \mathfrak{r}_1 \beta = \mathfrak{d}_0 \mathfrak{d}_1 \alpha' = \mathfrak{d}_0 \mathfrak{d}_1 \beta',$$

we have

$$(\beta' \circ \alpha') * (\beta \circ \alpha) = (\beta' * \beta) \circ (\alpha' * \alpha).$$

b) the identity map of $\mathfrak{C}_0$ and the map $\mathfrak{i}_1 \colon \mathfrak{C}_1 \to C_2$ define a functor from $\mathfrak{C}_{0,1}$ to $\mathfrak{C}_{0,2}$.

It is clear that the 2-category $\mathfrak{C}$ is determined once we know the diagram of sets

$$\mathfrak{C}_2 \underset{\mathfrak{r}_1}{\overset{\mathfrak{d}_1}{\rightrightarrows}} \!\!\xleftarrow{\mathfrak{i}_1}\! \mathfrak{C}_1 \underset{\mathfrak{r}_0}{\overset{\mathfrak{d}_0}{\rightrightarrows}} \!\!\xleftarrow{\mathfrak{i}_0}\! \mathfrak{C}_0$$

and the maps

$$\mu_{1,2} \colon \mathfrak{C}_2 \underset{\mathfrak{r}_1, \mathfrak{d}_1}{\times} \mathfrak{C}_2 \to \mathfrak{C}_2, \qquad \mu_{0,1} \colon \mathfrak{C}_1 \underset{\mathfrak{r}_0, \mathfrak{d}_0}{\times} \mathfrak{C}_1 \to \mathfrak{C}_1$$

$$\mu_{0,2} \colon \mathfrak{C}_2 \underset{\mathfrak{r}_0 \mathfrak{r}_1, \mathfrak{d}_0 \mathfrak{d}_1}{\times} \mathfrak{C}_2 \to \mathfrak{C}_2.$$

Besides, we can show (but we will not use this result) that these conditions define a 2-category if and only if the following three conditions are satisfied:

(i) $\mathfrak{r}_0 \mathfrak{r}_1 = \mathfrak{r}_0 \mathfrak{d}_1$ and $\mathfrak{d}_0 \mathfrak{r}_1 = \mathfrak{d}_0 \mathfrak{d}_1$.

(ii) $\mu_{1,2}$ (resp. $\mu_{0,1}$, $\mu_{0,2}$) is the composition map of a category whose set of objects is $\mathfrak{C}_1$ (resp. $\mathfrak{C}_0$, $\mathfrak{C}_0$), whose set of morphisms is $\mathfrak{C}_2$ (resp. $\mathfrak{C}_1$, $\mathfrak{C}_2$), whose domain and range maps are $\mathfrak{d}_1$ and $\mathfrak{r}_1$ (resp. $\mathfrak{d}_0$ and $\mathfrak{r}_0$, $\mathfrak{d}_0 \mathfrak{d}_1$ and $\mathfrak{r}_0 \mathfrak{r}_1$), and where $\mathfrak{i}_1$ (resp. $\mathfrak{i}_0$, $\mathfrak{i}_1 \mathfrak{i}_0$) is the map which associates with each object its identity morphism.

(iii) The relations a) and b) of 2.3 are satisfied.

1.4. From now on, we will consider only 2-categories $\mathfrak{X}$ which satisfy conditions A and B below:

A. *If x and y are two objects of $\mathfrak{X}$, $\mathcal{H}om_{\mathfrak{X}}(x, y)$ is a groupoid.*

B. *There is an object o of $\mathfrak{X}$ such that, for any object x of $\mathfrak{X}$, $\mathcal{H}om_{\mathfrak{X}}(o, x)$ and $\mathcal{H}om_{\mathfrak{X}}(x, o)$ are isomorphic to the zero category $\mathcal{O}$.*

Any object satisfying condition B will be called a zero object. Among all zero objects, we will choose one which will be written $o_{\mathfrak{X}}$ (or simply o), and we will say that $o_{\mathfrak{X}}$ is the (distinguished) zero object of $\mathfrak{X}$. For each object x of $\mathfrak{X}$, we will write 0^x (resp. 0_x) for the unique 1-morphism of domain (resp. range) x and range (resp. domain) $o_{\mathfrak{X}}$. If x and y are two objects of $\mathfrak{X}$, we write 0^x_y or symply 0 for the 1-morphism $0_y \cdot 0^x$ (it is the *zero* 1-morphism from x to y, which is independent of the choice of $o_{\mathfrak{X}}$).

In the sequel, we will constantly refer to the 2-category $\mathfrak{Gr}$ *of pointed groupoids*: and object of $\mathfrak{Gr}$ is a pointed groupoid, i.e. a pair (G, g_0) formed by a groupoid G (belonging to the universe $\mathcal{U}$ fixed once for all), and by an object g_0 of G, called *base point* or *indexed point*. If $x=(G, g_0)$ and $y=(H, h_0)$ are two pointed groupoids and if $\mathcal{H}om(G, H)$ is the category of all functors from G to H, we choose $\mathcal{H}om_{\mathfrak{Gr}}(x, y)$ to be the subcategory of $\mathcal{H}om(G, H)$ whose objects are the functors sending g_0 to h_0, whose morphisms are the functor homomorphisms φ such that $\varphi(x_0)=\operatorname{Id} y_0$. Finally, the composition functors of $\mathfrak{Gr}$ are the obvious functors.

1.5. If x and y are two objects of a 2-category $\mathfrak{X}$ which satisfies A and B, we will always give to $\mathcal{H}om_{\mathfrak{X}}(x, y)$ a base point by indexing the point 0^x_y. Besides, if z is a third object, we have a commutative diagram (Fig. 48)

$$\begin{array}{ccccc}
\mathcal{O} \times \mathcal{H}om_{\mathfrak{X}}(y, z) & \xrightarrow{0^x_y \times \operatorname{Id}} & \mathcal{H}om_{\mathfrak{X}}(x, y) \times \mathcal{H}om_{\mathfrak{X}}(y, z) & \xleftarrow{\operatorname{Id} \times 0^y_z} & \mathcal{H}om_{\mathfrak{X}}(x, y) \times \mathcal{O} \\
\downarrow & & \downarrow{\scriptstyle \mu_{x,y,z}} & & \downarrow \\
\mathcal{O} & \longrightarrow & \mathcal{H}om_{\mathfrak{X}}(x, z) & \longleftarrow & \mathcal{O}
\end{array}$$

Fig. 48

In particular, the composition functor sends the base point $(0^x_y, 0^y_z)$ of $\mathcal{H}om_{\mathfrak{X}}(x, y) \times \mathcal{H}om_{\mathfrak{X}}(y, z)$ to the base point of $\mathcal{H}om(x, z)$.

1.6. If x and y are two objects of $\mathfrak{X}$, we write $|x, y|$ for the pointed set $\Pi_0(\mathcal{H}om_{\mathfrak{X}}(x, y))$; the base point is obviously the connected component of 0^x_y. Since the set of connected components of a product of groupoids is identified with the product of the sets of connected components, the composition functors

$$\mu_{x,y,z}\colon \mathcal{H}om_{\mathfrak{X}}(x, y) \times \mathcal{H}om_{\mathfrak{X}}(y, z) \to \mathcal{H}om_{\mathfrak{X}}(x, z)$$

induce maps

$$|x, y, z| : |x, y| \times |y, z| \to |x, z|$$

and define a category $\overline{\mathfrak{X}}$ whose set of objects is $\mathfrak{X}_0$, and whose set of morphisms of domain x and range y is $|x, y|$, the composition maps being the maps $|x, y, z|$. If $f: x \to y$ is an object of $\mathcal{H}om_{\mathfrak{X}}(x, y)$, we will write $[f]$ for the image of f in $|x, y|$.

We will say that $\overline{\mathfrak{X}}$ is the category "$\mathfrak{X}$ modulo homotopy"; it is clear that $o_{\mathfrak{X}}$ is a zero object of this category.

2. Exact Sequences of Pointed Groupoids

Before seeing how we can construct exact sequences in certain 2-categories $\mathfrak{X}$ satisfying condidions A and B, we will first consider the "standard" 2-category $\mathfrak{Gr}$.

2.1. Let (X, x_0) and (Y, y_0) be two pointed groupoids, and $f: X \to Y$ a 1-morphism. Let us write Γf for the following pointed groupoid: an object of Γf is a pair (x, h) formed by an object x of X and a morphism $h: y_0 \to f(x)$ of Y; a morphism from (x, h) to (x', h') is a morphism $\xi: x \to x'$ of X such that $h' = (f\xi) \circ h$; the composition of these morphisms is induced by the composition in X; finally, the base point of Γf is the pair $(x_0, \mathrm{Id}\, y_0)$. We write $pf: \Gamma f \to X$ for the 1-morphism $(x, h) \rightsquigarrow x$ and $hf: 0 \to f(pf)$ for the 2-morphism — i.e. the functor homomorphism — which associates with each object (x, h) of Γf the morphism $h: y_0 \to f(x)$.

2.2. Given these definitions, it is clear that we can describe $\Gamma(pf)$ as follows: and object is a pair (h, h_1) formed by a morphism $h_1: x_0 \to x$ of X and a morphism $h: y_0 \to f(x)$ of Y; a morphism from (h, h_1) to (h', h_1') is a morphism ξ of X such that $h_1' = \xi \circ h_1$ and $h' = f(\xi) \circ h$; these morphisms are composed "as in X"; finally, the base point of $\Gamma(pf)$ is the pair $(\mathrm{Id}\, y_0, \mathrm{Id}\, x_0)$. The 1-morphism $p^2 f = p(pf)$ associates with (h, h_1) the pair (x, h), where x is the range of h_1; the 2-morphism $h(p f)$ associates with (h, h_1) the morphism $h_1: 0 \to (p f) \cdot (p^2 f)$.

If follows from the above description that two objects of $\Gamma(p f)$ are connected by at most one invertible morphism, i.e. that $\Gamma(p f)$ is equivalent to a discrete groupoid (see II). More precisely, according to the conventions of the beginning of the chapter, let us write $y_0: \mathcal{O} \to Y$ for the unique functor which sends the point of $\mathcal{O}$ to y_0, and let us write ΩY instead of Γy_0; this groupoid is discrete, and its set of objects is the Poincaré group $\Pi_1(Y, y_0)$; it is related to $\Gamma(p f)$ by means of the commutative triangle

$$\begin{array}{ccc} \Omega Y & \xrightarrow{rf} & \Gamma(pf) \\ & {\scriptstyle qf}\searrow \quad \swarrow{\scriptstyle p^2 f} & \\ & \Gamma f & \end{array}$$

where $q\,f$ and $r\,f$ are defined by the equalities

$$(q\,f)\,(l) = (x_0, l) \quad \text{and} \quad (r\,f)\,(l) = (l, \mathrm{Id}\,x_0).$$

Finally, let $r'f\colon \Gamma(pf) \to \Omega Y$ be the functor which sends (h, h_1) to $f(h_1)^{-1} \circ h$. It is clear that $[r'f]$ is the inverse of $[rf]$ in the category $\overline{\mathfrak{Gr}}$, in other words, that $r'f$ is a functor quasi-inverse to rf.

2.3. A 1-morphism $g\colon \Omega X \to \Omega X$ is obviously identified with a map from $\Pi_1(X, x_0)$ to $\Pi_1(X, x_0)$ which preserves the neutral element. In particular, we write

$$\sigma X\colon \Omega X \to \Omega X$$

or simply σ for the 1-morphism which induces the map $l \rightsquigarrow l^{-1}$ on $\Pi_1(X, x_0)$. With the notations of 2.2, we see then that the triangle

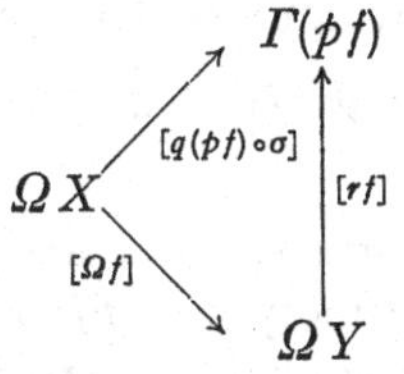

is commutative: here, Ωf is the 1-morphism which sends an object $\alpha\colon x_0 \to x_0$ of ΩX to $f(\alpha)\colon y_0 \to y_0$; similarly, $q(pf)$ sends α to the pair $(\mathrm{Id}\,y_0, \alpha)$, while $r(f) \cdot \sigma \cdot \Omega(f)$ sends α to $(f(\alpha^{-1}), \mathrm{Id}\,x_0)$; it is then quite clear that α is an isomorphism from $\big(f(\alpha^{-1}), \mathrm{Id}\,x_0\big)$ onto $(\mathrm{Id}\,y_0, \alpha)$, which proves the commutativity of the above diagram.

2.4. Putting the preceding statements one at the end of the other, we obtain a commutative diagram of $\overline{\mathfrak{Gr}}$ (Fig. 49)

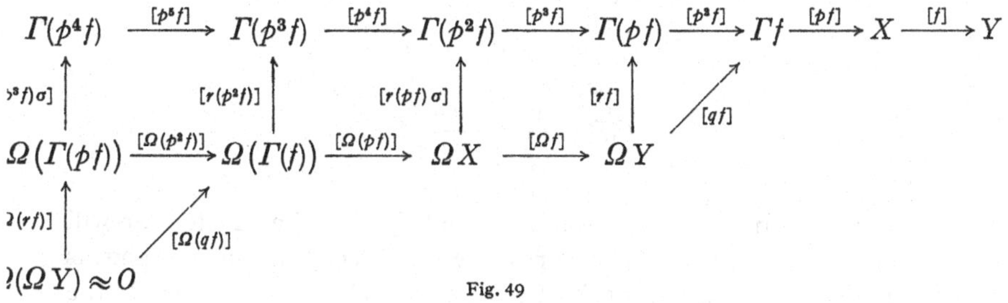

Fig. 49

where vertical arrows are isomorphisms.

However, if $g\colon G \to G'$ is a 1-morphism such that $[g]$ is invertible, then $\Pi_0(g)\colon \Pi_0(G) \to \Pi_0(G')$ is a bijection. Moreover, each 1-morphism $f\colon X \to Y$ induces an exact sequence

$$\Pi_0(\Gamma f) \xrightarrow{\Pi_0(pf)} \Pi_0(X) \xrightarrow{\Pi_0(f)} \Pi_0(Y)$$

of pointed sets, and $\Pi_0(\Omega Y)$ is identified with $\Pi_1(Y, y_0)$. Hence we can deduce from the above commutative diagram the following exact sequence of pointed sets:

$$(1) \to \Pi_1(\Gamma f) \xrightarrow{\Pi_1(\Gamma f)} \Pi_1(X) \xrightarrow{\Pi_1(f)} \Pi_1(Y) \xrightarrow{\Pi_0(qf)} \Pi_0(\Gamma f) \xrightarrow{\Pi_0(qf)} \Pi_0(X) \xrightarrow{\Pi_0(f)} \Pi_0(Y).$$

2.4.1. It is clear that

$$(1) \to \Pi_1(\Gamma f) \xrightarrow{\Pi_1(pf)} \Pi_1(X) \xrightarrow{\Pi_1(f)} \Pi_1(Y)$$

is an exact sequence of groups. But we have, moreover, that $\Pi_1(Y)$ *right-operates on* $\Pi_0(\Gamma f)$ as follows: if (x, h) is an element of Γf, and if α belongs to $\Pi_0(\Omega Y) = \Pi_1(X)$, we set $(x, h)\alpha = (x, h\alpha)$. Passing to connected components, we define the operation we were looking for.

3. Spaces of Loops

Let us return now to the general case of a 2-category $\mathfrak{X}$ satisfying conditions A and B of 1.4.

3.1. Let $f\colon x \to y$ be a morphism of $\mathfrak{X}$ and $\Phi f = (\Gamma f, pf, hf)$ a triple formed by an object Γf, a 1-morphism $pf\colon \Gamma f \to x$, and a 2-morphism $hf\colon 0_y^{\Gamma f} \to f \cdot (pf)$. For each object t of $\mathfrak{X}$, the triple Φf determines a functor

$$\varphi^t f\colon (\Gamma f)^t \to \Gamma(f^t)$$

(see 1.1 for the notations $(\Gamma f)^t$ and f^t, and 2.1 for the definition of $\Gamma(f^t)$). This functor associates with each 1-morphism $g\colon t \to \Gamma f$ the pair $((pf)g, (hf) * g)$ which is an object of $\Gamma(f^t)$, and with each 2-morphism $\gamma\colon g \to g'$ the 2-morphism $(pf) * \gamma$ from $((pf)g, (hf) * g)$ to $((pf)g', (hf) * g')$ (see 1.3 for the notation $*$). It follows from these definitions that we have a commutative diagram of pointed groupoids (Fig. 50)

$$\begin{array}{ccccc}
(\Gamma f)^t & \xrightarrow{(pf)^t} & x^t & \xrightarrow{f^t} & y^t \\
\big\downarrow{\scriptstyle \varphi^t f} & & \big\downarrow{\scriptstyle \mathrm{Id}} & & \big\downarrow{\scriptstyle \mathrm{Id}} \\
\Gamma(f^t) & \xrightarrow{p(f^t)} & x^t & \xrightarrow{f^t} & y^t
\end{array}$$

Fig. 50

3.2. If X and Y are two groupoids, and $F\colon X \to Y$ a functor, we will say that F is *connected* if F is surjective on objects and on morphisms and if, for each object b of Y, the groupoid $F^{-1}(b)$ is connected (the objects of the groupoid $F^{-1}(b)$ are the objects of X which are sent to b, and its morphisms are the morphisms of X which are sent to $\mathrm{Id}\, b$), Such a functor obviously induces a bijection from $\Pi_0(X)$ onto $\Pi_0(Y)$.

We are now ready to make another hypothesis C on our 2-category $\mathfrak{X}$.

C. *For each* 1-*morphism* $f\colon x \to y$ *of* $\mathfrak{X}$, *there is a triple* $(\Gamma f, pf, hf)$ *such that the functor* $\varphi^t f\colon (\Gamma f)^t \to \Gamma(f^t)$ *is connected for each object* t *of* $\mathfrak{X}$.

If we go back to the definitions, we see that the *connectivity* condition is the conjunction of conditions C_1 and C_2.

C_1. For each 1-morphism $g\colon t \to x$ and each 2-morphism $h\colon 0 \to fg$ there is a 1-morphism $g'\colon t \to \Gamma f$ such that $g = (pf)g'$ and $h = (hf) * g'$. Moreover, the relations $g = (pf)g' = (pf)g''$ and $(hf) * g' = (hf) * g''$ imply the existence of a 2-morphism $\gamma\colon g' \to g''$ such that $(pf) * \gamma = \operatorname{Id} g$.

C_2. If $\gamma\colon (pf)g' \to (pf)g''$ is a 2-morphism such that $(f * \gamma) \circ \big((hf) * g'\big) = (hf) * g''$, there is a 2-morphism $\delta\colon g' \to g''$ such that $\gamma = (pf) * \delta$.

3.2.1. If f is the zero morphism $0\colon o \to y$ and if the triple $(\Gamma f, pf, hf)$ satisfies C, we will write Ωy instead of Γf, and we will say that Ωy is a *space of loops of* y.

3.2.2. The 2-category $\mathfrak{Gr}$ of groupoids satisfies condition C: it is sufficient to define $(\Gamma f, pf, hf)$, as done in 2.1. Moreover, with the notations of 3.2, C_1, the 1-morphism g' is determined in a unique way by g and h. Similarly, in 3.2, C_2, δ is determined by γ.

3.3. From now on, we will suppose that the 2-category $\mathfrak{X}$ satisfies conditions A, B, C given above.

For each 1-morphism $f\colon x \to y$ of $\mathfrak{X}$, each triple $\Phi f = (\Gamma f, pf, hf)$ satisfying C, and each object t, the functor $\varphi^t f$ of 3.1 induces a bijection from $\Pi_0 (\Gamma f)^t$ onto $\Pi_0 \Gamma(f^t)$:

$$[\varphi^t f]\colon |t, \Gamma f| \to \Pi_0 \Gamma(f^t).$$

This implies that the functor $t \rightsquigarrow \Pi_0 \Gamma(f^t)$ from $(\overline{\mathfrak{X}})^\circ$ to $\mathscr{E}$ is representable by Γf; more precisely, $\big(\Gamma f, (pf, hf)\big)$ is a representation of this functor, so that Γf and $[pf]$ are determined up to an isomorphism of $\overline{\mathfrak{X}}$.

Because of this unicity "up to homotopy equivalence", from now on, we will associate with each 1-morphism $f\colon x \to y$ a triple Φf chosen arbitrarily among all those satisfying condition C. We will see in the sequel in what measure the constructions based on this association depend on the choice of Φf.

In the particular case where f is the zero 1-morphism $o \to y$, Γf is the space of loops of y (3.2.1) and represents the functor from $(\overline{\mathfrak{X}})^\circ$ to $\mathscr{E}$ which associates with each t the Poincaré group $\Pi_1(y^t, 0)$ of $\mathscr{H}om_{\mathfrak{X}}(t, y)$ at the point 0_y^t. It follows obviously that *Ωy is given a natural group structure in the category $\overline{\mathfrak{X}}$.*

3.4. In order to study how Γf depends on the 1-morphism f, we will introduce the following category $\mathscr{A}r\,\mathfrak{X}$: the objects of $\mathscr{A}r\,\mathfrak{X}$ are the 1-morphisms of $\mathfrak{X}$; its morphisms are the diagrams

$$(\chi)\qquad \begin{array}{ccc} x & \xrightarrow{\ f\ } & y \\ {\scriptstyle u}\big\downarrow & \overset{\alpha}{\swarrow} & \big\downarrow{\scriptstyle v} \\ x' & \xrightarrow[\ f'\ ]{} & y' \end{array}$$

where f, f', u, v are 1-morphisms and $\alpha\colon vf \to f'u$ a 2-morphism. The domain (resp. range) of χ is f (resp. f'). Finally, we define the composition $\chi''=\chi'\circ\chi$ of the two diagrams

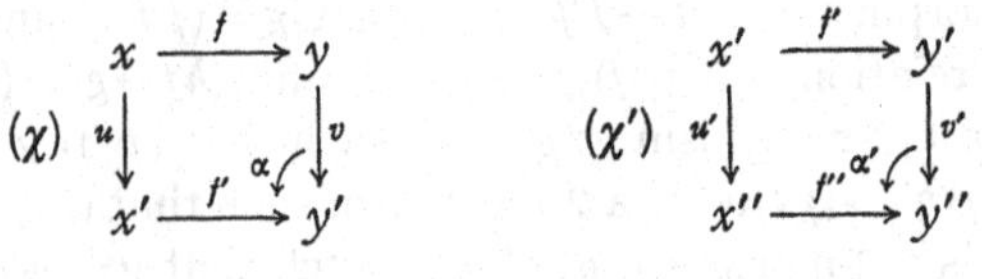

to be the diagram

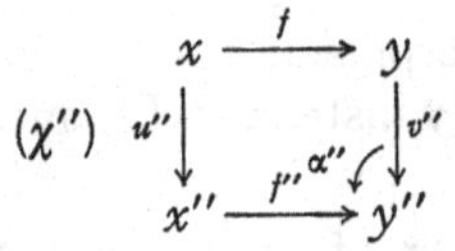

where $u''=u'u$, $v''=v'v$, and $\alpha''=(\alpha' * u)\circ(v' * \alpha)$.

Let then χ be a morphism of $\mathscr{A}r\,\mathfrak{X}$. By condition C_1 of 3.2, there is a 1-morphism $\Gamma\chi\colon \Gamma f \to \Gamma f'$ such that $(pf')(\Gamma\chi)=u(pf)$ and that $(hf') * (\Gamma\chi)$ is the composed 2-morphism

$$0 \xrightarrow{v*(hf)} vf(pf) \xrightarrow{\alpha*(pf)} f'u(pf).$$

The class $[\Gamma\chi]$ of such a 1-morphism is uniquely determined by χ (3.3).

3.4.1. In the particular case where $\mathfrak{X}$ is the 2-category of pointed groupoids, the 1-morphism $\Gamma\chi$ is also uniquely determined by the conditions of 3.4; it is the functor $(x,h) \rightsquigarrow (ux, (\alpha x)\circ(vh))$ from Γf to $\Gamma f'$.

Moreover, when u and v are equivalences of categories, we see easily that $\Gamma\chi$ is an equivalence, or that $[\Gamma\chi]$ is an isomorphism of $\overline{\mathfrak{Gr}}$.

3.4.2. By 3.4.1, we see also that $\Gamma(\chi'\circ\chi)=\Gamma(\chi)\circ\Gamma(\chi')$ when χ' and χ are two composable diagrams of $\mathfrak{Gr}$. More generally, if $\mathfrak{X}$ is a 2-category satisfying conditions A, B and C, the definitions imply the commutativity of the diagrams

(*)

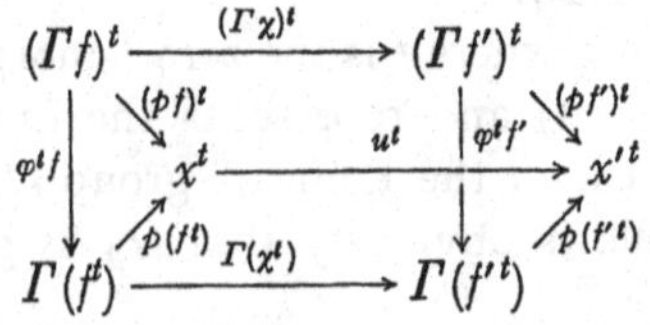

Fig. 51

of $\mathfrak{Gr}$ associated with a morphism χ of $\mathscr{A}r\,\mathfrak{X}$. If we identify $|t,\Gamma f|$ and $|t,\Gamma f'|$ with $\Pi_0\Gamma(f^t)$ and $\Pi_0\Gamma(f'^t)$ by means of $\Pi_0\varphi^t f$ and $\Pi_0\varphi^t f'$, we see that $|t,\Gamma\chi|$ is identified with $\Pi_0(\Gamma\chi^t)$. In particular, if u and v are homotopy equivalences, i.e. if $[u]$ and $[v]$ are invertible in $\overline{\mathfrak{X}}$, u^t and v^t are equivalences of categories for all t, and hence $\Pi_0(\Gamma\chi^t)$ and $|t,\Gamma\chi|$ are bijections: *if* $[u]$ *and* $[v]$ *are invertible, then* $[\Gamma\chi]$ *is invertible.*

Finally, if (χ) and (χ') are two composable morphisms of $\mathcal{A}r\,\mathfrak{X}$ and if $\chi''=\chi'\circ\chi$, we have $\chi''^{t}=(\chi'^{t})\circ(\chi^{t})$, and hence $\Gamma(\chi''^{t})=\Gamma(\chi'^{t})\circ\Gamma(\chi^{t})$. Then we have the equality

$$[\Gamma(\chi'\circ\chi)]=[\Gamma\chi']\circ[\Gamma\chi]$$

which show that Γ is a functor from $\mathcal{A}r\,\mathfrak{X}$ to $\overline{\mathfrak{X}}$.

3.4.3. In the particular case where χ is the commutative diagram

$$(\chi)\quad \begin{array}{ccc} 0 & \longrightarrow & x \\ \big\downarrow & \mathrm{Id}\swarrow & \big\downarrow{\scriptstyle f} \\ 0 & \longrightarrow & y \end{array}$$

$\Gamma\chi$ is a 1-morphism $\Omega x\to\Omega y$. We write then Ωf instead of $\Gamma\chi$, and *we see that Ω is a functor from $\mathfrak{X}_{0,1}$ to $\overline{\mathfrak{X}}$* (1.3).

In the particular case where $\mathfrak{X}$ the 2-category $\mathfrak{Gr}$ of pointed groupoids we see that for each 1-morphism $F\colon X\to Y$, $\Pi_0(\Omega F)$ is a morphism of groups. If we apply diagram $(*)$ of 3.4.2 to the morphism χ considered here (Fig. 52)

$$\begin{array}{ccc} (\Omega x)^{t} & \xrightarrow{(\Omega f)^{t}} & (\Omega y)^{t} \\ {\scriptstyle \varphi^{t}0_x}\big\downarrow & & \big\downarrow{\scriptstyle \varphi^{t}0_y} \\ \Omega(x^{t}) & \xrightarrow{\Omega(f^{t})} & \Omega(y^{t}) \end{array}$$

Fig. 52

and if we note that $[\varphi^{t}0_x]$, $[\varphi^{t}0_y]$ give to Ωx and Ωy a group structure in the category $\overline{\mathfrak{X}}$, we see then that $[\Omega f]$ is a *homomorphism of $\overline{\mathfrak{X}}$-groups.* Finally, Ω can be considered as *a functor from $\mathfrak{X}_{0,1}$ to the category of $\overline{\mathfrak{X}}$-groups.*

Moreover, if f and f' are two 1-morphisms from x to y which can be connected by a 2-morphism $\alpha\colon f\to f'$, $\Omega(f^{t})$ and $\Omega(f'^{t})$ coincide for all t. Hence the same holds for $|t,\Omega f|$ and $|t,\Omega f'|$, i.e. for $[\Omega f]$ and $[\Omega f']$. Hence the equality $[f]=[f']$ implies $[\Omega f]=[\Omega f']$, so that Ω *induces a functor* (still written Ω!) *from* $\overline{\mathfrak{X}}$ to $\overline{\mathfrak{X}}$.

3.4.4. Finally, if we take χ to be the diagram

$$\begin{array}{ccc} 0 & \longrightarrow & y \\ \big\downarrow & \mathrm{Id}\swarrow & \big\downarrow{\scriptstyle \mathrm{Id}} \\ x & \xrightarrow{\ f\ } & y \end{array}$$

Γx is a 1-morphism $qf\colon \Omega y\to\Gamma f$, whose class qf is uniquely determined by the conditions $(pf)(qf)=0_x^{\Omega y}$ and $(hf)*(qf)=h(0_y)$ (see 3.4).

4. Exact Sequences: Statement of the Theorem and Invariance

A composable sequence of $\overline{\mathfrak{X}}$ *of length* n (or simply a sequence of $\overline{\mathfrak{X}}$ of length n) is a diagram of $\overline{\mathfrak{X}}$ of the form

$$x_n \xrightarrow{f_n} x_{n-1} \cdots \longrightarrow x_2 \xrightarrow{f_2} x_1 \xrightarrow{f_1} x_0 .$$

If

$$y_n \xrightarrow{g_n} y_{n-1} \cdots \longrightarrow y_2 \xrightarrow{g_2} y_1 \xrightarrow{g_1} y_0$$

is another sequence of length n, a morphism from the first to the second is, by definition, a sequence of morphisms $h_i\colon x_i \to y_i$ of $\overline{\mathfrak{X}}$ such that $h_i f_{i+1} = g_{i+1} h_{i+1}$ for $0 \leqq i \leqq n-1$. These morphisms are composed in the obvious way, so that we will be able to speak of *the category of composable sequences of* $\overline{\mathfrak{X}}$ *of length* n.

4.1. Let us call *selection* any pair $\mathscr{S} = (\Phi, \Psi)$ formed by two maps Φ and Ψ which satisfy the following conditions: the map Φ associates with each object f of $\mathscr{A}r\,\mathfrak{X}$ a triple $(\Gamma f, pf, hf)$ satisfying condition C of 3.2; the map associates with each morphism χ of $\mathscr{A}r\,\mathfrak{X}$ a 1-morphism $\Gamma\chi$ of $\mathfrak{X}$ satisfying the conditions of 3.4.

With each selection $\mathscr{S}$ *is canonically associated a functor* $\Omega^{\mathscr{S}}$ *from* $\mathscr{A}r\,\mathfrak{X}$ *to the category of composable sequences of* $\overline{\mathfrak{X}}$ *of infinite length*: this functor associates with an object $f\colon x \to y$ the infinite sequence

$$\cdots \Omega^2 x \xrightarrow{[\Omega^2 f]} \Omega^2 y \xrightarrow{[\Omega q f]} \Omega\Gamma f \xrightarrow{[\Omega p f]} \Omega x \xrightarrow{[\Omega f]} \Omega y \xrightarrow{[qf]} \Gamma f \xrightarrow{[pf]} x \xrightarrow{[f]} y$$

defined with the help of 3.4.3 and 3.4.4. It associates with a morphism

$$(\chi)\qquad \begin{array}{ccc} x & \xrightarrow{f} & y \\ {\scriptstyle u}\downarrow & \overset{\alpha}{\swarrow} & \downarrow{\scriptstyle v} \\ x' & \xrightarrow{f'} & y' \end{array}$$

of $\mathscr{A}r\,\mathfrak{X}$ the following morphism of composable sequences of $\overline{\mathfrak{X}}$ (Fig. 53)

$$\begin{array}{ccccccccccccc} \vdots\ \Omega^2 y & \xrightarrow{[\Omega q f]} & \Omega\Gamma f & \xrightarrow{[\Omega p f]} & \Omega x & \xrightarrow{[\Omega f]} & \Omega y & \xrightarrow{[qf]} & \Gamma f & \xrightarrow{[pf]} & x & \xrightarrow{[f]} & y \\ \downarrow{\scriptstyle [\Omega^2 v]} & & \downarrow{\scriptstyle [\Omega\Gamma\chi]} & & \downarrow{\scriptstyle [\Omega u]} & & \downarrow{\scriptstyle [\Omega v]} & & \downarrow{\scriptstyle [\Gamma\chi]} & & \downarrow{\scriptstyle [u]} & & \downarrow{\scriptstyle [v]} \\ \vdots\ \Omega^2 y' & \xrightarrow{[\Omega q f']} & \Omega\Gamma f' & \xrightarrow{[\Omega p f']} & \Omega x' & \xrightarrow{[\Omega f']} & \Omega y' & \xrightarrow{[qf']} & \Gamma f' & \xrightarrow{[pf']} & x' & \xrightarrow{[f']} & y' \end{array}$$

Fig. 53

The equality $[u][pf] = [pf'][\Gamma\chi]$ follows directly from the definition of $\Gamma\chi$. If we show that we also have $[\Gamma\chi][qf] = [qf'][\Omega v]$, then the commutativity of the other squares will follow from the last part of 3.4.3: now by 3.4.2 and 3.4.4, $[\Gamma\chi][qf]$ and $[qf'][\Omega v]$ are identified with $[\Gamma\chi']$ and $[\Gamma\chi'']$, where χ' and χ'' denote respectively the compositions of Fig. 54

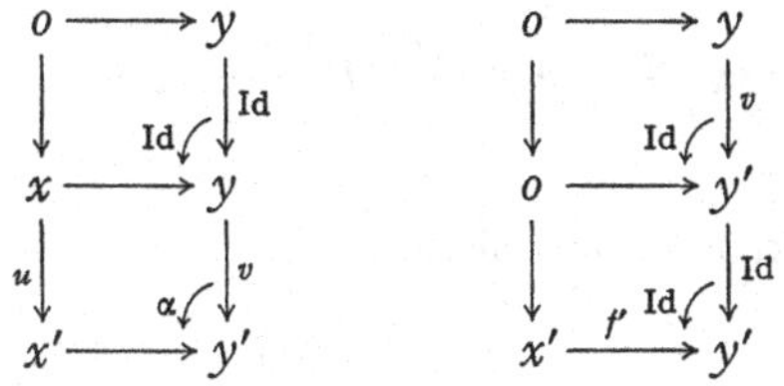

Fig. 54

Thus our statement follows from the equality $\chi'=\chi''$.

4.2. *Theorem: Let $\mathfrak{X}$ be a 2-category satisfying conditions A, B and C' and $f: x \to y$, a 1-morphism of $\mathfrak{X}$. For each object t of $\mathfrak{X}$, the sequence*

$$\ldots |t, \Omega x| \xrightarrow{|t, \Omega f|} |t, \Omega y| \xrightarrow{|t, qf|} |t, \Gamma f| \xrightarrow{|t, pf|} |t, x| \xrightarrow{|t, f|} |t, y|$$

is an exact sequence of pointed sets.

Recall that a sequence

$$T_n \xrightarrow{\tau_n} T_{n-1} \cdots \longrightarrow T_1 \xrightarrow{\tau_1} T_0$$

of pointed sets is exact if, for $1 \leq i \leq n-1$, $\tau_{i+1}(T_{i+1})$ is the inverse image of the base point of T_{i-1}, under τ_i. In the exact sequence of the theorem, the pointed sets $|t, \Omega y|$, $|t, \Omega x|$, ..., are groups, the maps by which they are connected are group homomorphisms, and $|t, \Omega y|$ right-operates on $|t, \Gamma f|$ in a way which is compatible with $|t, qf|$ and with the operation of $|t, \Omega y|$ on itself defined by right translations.

From now on, we will say that a sequence

$$x_n \xrightarrow{f_n} x_{n-1} \cdots x_1 \xrightarrow{f_1} x_0$$

of $\overline{\mathfrak{X}}$ is *r-exact* if, for each object t of $\overline{\mathfrak{X}}$, the sequence of pointed sets

$$|t, x_n| \xrightarrow{t, f_n} |t, x_{n-1}| \ldots |t, x_1| \xrightarrow{t, f_1} |t, x_0|$$

is exact. With this terminology, the sequence $\Omega^{\mathcal{S}} f$ (4.1) is r-exact.

The proof of the theorem will be given in paragraph 5.

4.3. 1*st invariance Theorem: Let $f: x \to y$ and $f': x' \to y'$ be two 1-morphisms of $\mathfrak{X}$ such that $[f]$ and $[f']$ are isomorphic morphisms of $\overline{\mathfrak{X}}$* (hence there are 1-morphisms $u\colon x\to x'$ and $v\colon y\to y'$ such that $[u]$ and $[v]$ are invertible and satisfy the equation $[f'][u]=[v][f]$). *In that case, the r-exact sequences*

$$\ldots\Omega x \xrightarrow{[\Omega f]} \Omega y \xrightarrow{[qf]} \Gamma f \xrightarrow{[pf]} x \xrightarrow{[f]} y$$

and

$$\ldots\Omega x' \xrightarrow{[\Omega f']} \Omega y' \xrightarrow{[qf']} \Gamma f' \xrightarrow{[pf']} x' \xrightarrow{[f']} y'$$

of $\overline{\mathfrak{X}}$ *are isomorphic.*

Indeed, there exists a morphism

$$(\chi)\qquad \begin{array}{ccc} x & \xrightarrow{f} & y \\ {\scriptstyle u}\downarrow & \overset{\alpha}{\swarrow} & \downarrow{\scriptstyle v} \\ x' & \xrightarrow{f'} & y' \end{array}$$

of $\mathscr{A}r\,\mathfrak{X}$ such that $[u]$ and $[v]$ are invertible. By 3.4.2 and 3.4.3, $\Omega^{\mathscr{S}}\chi$ is an isomorphism from $\Omega^{\mathscr{S}}f$ onto $\Omega^{\mathscr{S}}f'$ (4.1).

Theorem 4.3 can be applied, in particular, to the case where $[f]=[f']$. Hence, *the r-exact sequence* $\Omega^{\mathscr{S}}f$ *depends only on the class* $[f]$ *of* f, *up to isomorphism.*

4.4. The functor

$$\Omega^{\mathscr{S}}\colon f \rightsquigarrow (\ldots\Omega y \xrightarrow{[qf]} \Gamma f \xrightarrow{[pf]} x \xrightarrow{[f]} y)$$

obviously depends on the choice we made for $\mathscr{S}$. However, this dependence is very "loose":

2nd invariance Theorem: Let $\mathscr{S}$ *and* $\mathscr{S}'$ *be two selections relatively to the same 2-category* $\mathfrak{X}$ *which satisfies A, B and C. Then the functors* $\Omega^{\mathscr{S}}$ *and* $\Omega^{\mathscr{S}'}$ *are isomorphic.*

Let us write $\Gamma' f, \Omega' y, \Omega' f, p' f, h' f, q' f, \ldots$ for the items corresponding to $\Gamma f, \Omega y, \Omega f, pf, hf, qf, \ldots$ but constructed from the selection $\mathscr{S}'$. Let us associate with each 1-morphism $f\colon x\to y$ a 1-morphism $wf\colon \Gamma f\to\Gamma' f$ such that $(p' f)(wf)=pf$ and $(h' f)*(wf)=hf$. For each object t of $\mathfrak{X}$, we then have the equality $(\varphi'^t f)(wf)^t=\varphi^t f$ (notation of 3.1). We deduce that $[wf]$ is invertible.

Moreover, consider a morphism

$$(\chi)\qquad \begin{array}{ccc} x & \xrightarrow{f} & y \\ {\scriptstyle u}\downarrow & \overset{\alpha}{\swarrow} & \downarrow{\scriptstyle v} \\ x' & \xrightarrow{f'} & y' \end{array}$$

of $\mathscr{A}r\,\mathfrak{X}$. We then have equalities $(p' f')(wf')(\Gamma\chi)=(pf')(\Gamma\chi)=u(pf)=u(p' f)(wf)=(p' f')(\Gamma'\chi)(wf)$ and $(h' f')*(wf')*(\Gamma\chi)=(hf')*(\Gamma\chi)=$

$(\alpha * pf) \circ (v * hf) = (\alpha * p'f * wf) \circ (v * h'f * wf) = [(\alpha * p'f) \circ (v * h'f)] * wf = (h'f') * (\Gamma'\chi) * (wf)$.

By 3.2, C_1, there is then a 2-morphism $\gamma: (wf')(\Gamma\chi) \to (\Gamma'\chi)(wf)$ such that $(p'f') * \gamma = \mathrm{Id}$:

$$\begin{array}{ccc} \Gamma f & \xrightarrow{\Gamma\chi} & \Gamma f' \\ {\scriptstyle wf}\downarrow & & \downarrow{\scriptstyle wf'} \\ \Gamma' f & \xrightarrow[\Gamma'\chi]{\ \gamma\ } & \Gamma' f' \end{array}$$

We have in particular $[wf'][\Gamma\chi] = [\Gamma'\chi][wf]$. Going back to the definitions, we easily deduce the commutativity of the diagram of Fig. 55

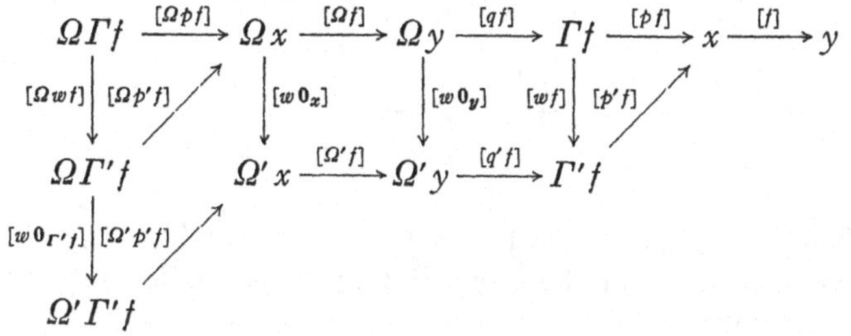

Fig. 55

Hence there is an isomorphism $Wf: \Omega^{\mathscr{S}} f \simeq \Omega^{\mathscr{S}'} f$ whose components are the morphisms $\mathrm{Id}\, y$, $\mathrm{Id}\, x$, $[wf]$, $[w0_y]$, $[w0_x]$, $[w0_{\Gamma' f}]\,[\Omega w f]$, $[w0_{\Omega' y}]\,[\Omega w 0_y]$, $[w0_{\Omega' x}]\,[\Omega w 0_x]$, $[w0_{\Omega'\Gamma' f}]\,[\Omega w 0_{\Gamma' f}]\,[\Omega^2 w f] \ldots$.

It remains to be proved that the Wf form a functor homomorphism from $\Omega^{\mathscr{S}}$ to $\Omega^{\mathscr{S}'}$. This is left to the reader.

5. Proof of Theorem 4.2

By 3.1 and 3.2, the sequence

$$|t, \Gamma f| \xrightarrow{|t, pf|} |t, x| \xrightarrow{|t, f|} |t, y|$$

is isomorphic to the sequence

$$\Pi_0 \Gamma(f^t) \xrightarrow{\Pi_0 p(f^t)} \Pi_0(x^t) \xrightarrow{\Pi_0(f^t)} \Pi_0(y^t).$$

It is then exact (2.4), and the same holds for the infinite sequence

$$\ldots |t, \Gamma p^2 f| \xrightarrow{|t, p^3 f|} |t, \Gamma p f| \xrightarrow{|t, p^2 f|} |t, \Gamma f| \xrightarrow{|t, pf|} |t, x| \xrightarrow{|t, f|} |t, y|$$

where $p^{n+1} f = p(p^n f)$.

We will show that this infinite sequence is isomorphic to the sequence of theorem 4.2; this will prove the exactness of the latter:

5.1. Let us associate with each 1-morphism $f: x \to y$ of $\mathfrak{X}$ a 1-morphism $rf: \Omega y \to \Gamma p f$ such that $(p^2 f)(rf) = qf$ and $(hpf) * (rf) = \mathrm{Id}\, 0_x^{\Omega y}$. When $\mathfrak{X}$ is the 2-category of pointed groupoids, rf is the functor considered in 2.2.

Similarly, let us associate with each object x of $\mathfrak{X}$ a 1-morphism σx (or simply σ) from Ωx to Ωx such that $(p 0_x)(\sigma x) = 0^{\Omega x}$ and $(h 0_x) * (\sigma x) = (h 0_x)^{-1}$. If $\mathfrak{X}$ is the 2-category of pointed groupoids, σx has already been defined in 2.3. In the general case, the definitions imply, for each object t, the commutativity of the square of Fig. 56

$$\begin{array}{ccc} (\Omega x)^t & \xrightarrow{(\sigma x)^t} & (\Omega x)^t \\ \downarrow{\scriptstyle \varphi^t 0_x} & & \downarrow{\scriptstyle \varphi^t 0_x} \\ \Omega(x^t) & \xrightarrow{\sigma(x^t)} & \Omega(x^t) \end{array}$$

Fig. 56

Since $\Pi_0 \sigma(x^t)$ is the endomorphism $\xi \rightsquigarrow \xi^{-1}$ of $\Pi_1(x^t, 0)$, $|t, \sigma x|$ is also the "inversion map" of the group $|t, \Omega x|$. This implies, in particular, that $[\sigma x]$ is invertible.

The morphisms rf and σx allow us to construct the following diagram of $\overline{\mathfrak{X}}$ (Fig. 57)

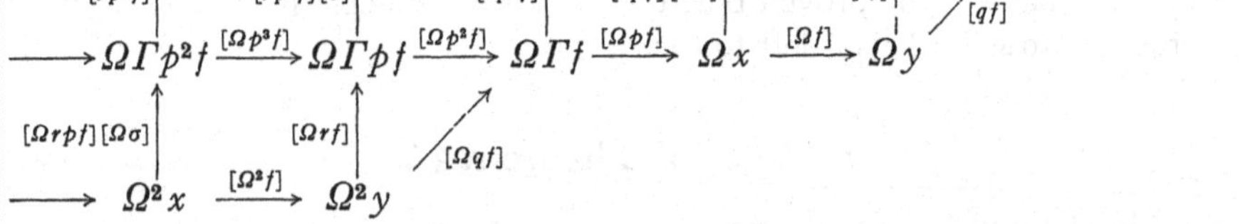

Fig. 57

We will show that this diagram is commutative and that the vertical arrows are invertible; this will prove Theorem 4.2. We obviously have $[p^2 f][rf] = [qf]$, by definition of rf. It will then be sufficient to prove that $[rf]$ is invertible and that $[rf][\Omega f] = [p^3 f][rpf][\sigma x]$ for each 1-morphism $f: x \to y$. In order to do this, we will construct the diagram reproduced on the next page. The lateral faces of this diagram will be commutative, and the vertical arrows will induce bijections on connected components. Our figure will then allow us to compare the top of the diagram to the bottom: thus we reduce the general case to the particular of pointed groupoids, which we considered in paragraph 2.

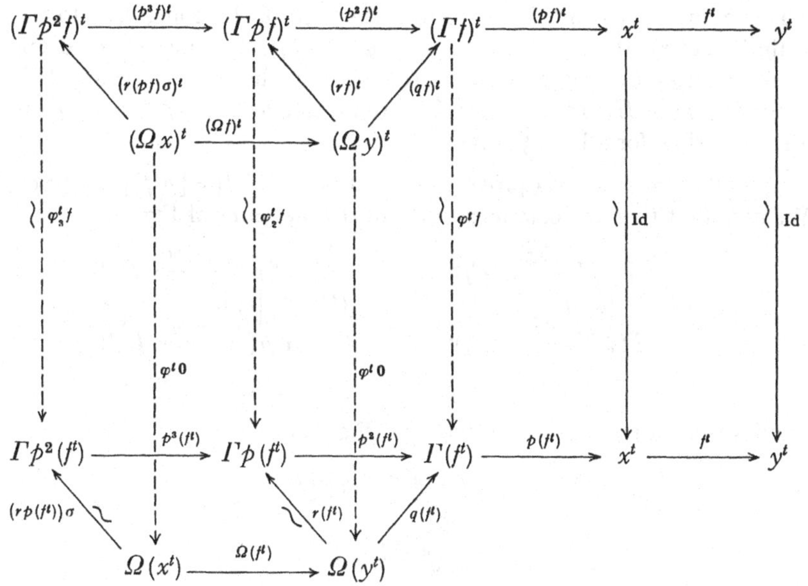

Fig. 58

Sketch intended to make the lecture of paragraph 5 easier.

5.2. Consider first $[rf]$:

Lemma: Let

$$(T)\qquad \begin{array}{ccc} X & \xrightarrow{F} & Y \\ {\scriptstyle T}\downarrow & & \downarrow{\scriptstyle \mathrm{Id}} \\ X' & \xrightarrow{F'} & Y \end{array}$$

be a commutative square of pointed groupoids. If T is connected, the functor $\Gamma T \colon \Gamma F \to \Gamma F'$ is also connected.

This lemma follows immediately from the definitions: it shows that the functor $\Gamma\varphi^t f \colon \Gamma(pf)^t \to \Gamma p(f^t)$, induced by the commutative square (Fig. 59)

$$(Tf)\qquad \begin{array}{ccc} (\Gamma f)^t & \xrightarrow{p(f)^t} & x^t \\ \downarrow{\scriptstyle \varphi^t f} & & \downarrow{\scriptstyle \mathrm{Id}} \\ \Gamma(f^t) & \xrightarrow{(pf^t)} & x^t \end{array}$$

Fig. 59

defined in 3.1, is connected. This functor can be inscribed in a commutative diagram (Fig. 60)

$$(Sf)\qquad \begin{array}{ccccc} (\Omega y)^t & & \xrightarrow{\quad \varphi^t 0_y \quad} & & \Omega(y^t) \\ \downarrow{\scriptstyle (rf)^t} & & & & \downarrow{\scriptstyle r(f^t)} \\ (\Gamma p f)^t & \xrightarrow{\varphi^t(pf)} & \Gamma(pf)^t & \xrightarrow{\Gamma\varphi^t f} & \Gamma p(f^t) \end{array}$$

Fig. 60

The commutativity of this diagram can be vertified by going back to the definition of rf. If we set then $\varphi_2^t f = (\Gamma\varphi^t f)(\varphi^t p f)$, the functor we obtain is the composition of two connected functor; hence it is connected. Since $\Pi_0(\varphi^t 0_y)$, $\Pi_0 r(f^t)$ and $\Pi_0(\varphi_2^t f)$ are invertible, $\Pi_0(rf)^t$ (i.e. $|t, rf|$) is also invertible for all t. Q.E.D.

5.3. We have now to prove the equality $[rf][\Omega f] = [p^3 f][rpf][\sigma x]$. We note first that the commutativity of the diagrams of Fig. 61

$$
(Tpf)\quad
\begin{array}{ccc}
(\Gamma p f)^t & \xrightarrow{(p^2 f)^t} & (\Gamma f)^t \\
\Big\downarrow{\scriptstyle \varphi^t(pf)} & & \Big\downarrow{\scriptstyle \mathrm{Id}} \\
\Gamma(pf)^t & \xrightarrow{p(pf)^t} & (\Gamma f)^t
\end{array}
\qquad\qquad
(U)\quad
\begin{array}{ccc}
\Gamma(pf)^t & \xrightarrow{p(pf)^t} & (\Gamma f)^t \\
\Big\downarrow{\scriptstyle \Gamma(\varphi^t f)} & & \Big\downarrow{\scriptstyle \varphi^t f} \\
\Gamma p(f^t) & \xrightarrow{p^2(f^t)} & \Gamma(f^t)
\end{array}
$$

Fig. 61

implies that of the composed diagram (Fig. 62)

$$
(\Xi)\quad
\begin{array}{ccc}
(\Gamma p f)^t & \xrightarrow{(p^2 f)^t} & (\Gamma f)^t \\
\Big\downarrow{\scriptstyle \varphi_2^t f} & & \Big\downarrow{\scriptstyle \varphi^t f} \\
\Gamma p(f^t) & \xrightarrow{p^2(f^t)} & \Gamma(f^t)
\end{array}
$$

Fig. 62

Moreover, since $[\varphi_2^t f]$ and $[\varphi^t f]$ are invertible by above, the same holds for $[\Gamma\Xi]$, where $\Gamma\Xi$ is the functor from $\Gamma(p^2 f)^t$ to $\Gamma p^2(f^t)$ defined by Ξ (3.4.2). The functor $\Gamma\Xi$ can be inscribed in the diagram of pointed groupoids

$$
\begin{array}{ccc}
(\Gamma p^2 f)^t & \xrightarrow{p^3(f)^t} & (\Gamma p f)^t \\
\Big\downarrow{\scriptstyle \varphi^t(p^2 f)} & \nearrow{\scriptstyle p(p^2 f)^t} & \Big\downarrow \\
\Gamma(p^2 f)^t & & {\scriptstyle \varphi_2^t f} \\
\Big\downarrow{\scriptstyle \Gamma\Xi} & & \Big\downarrow \\
\Gamma p^2(f^t) & \xrightarrow{(p^3 f^t)} & \Gamma p(f^t)
\end{array}
$$

Fig. 63

If we write $\varphi_3^t f = (\Gamma\Xi)\varphi^t(p^2 f)$, *$\varphi_3^t f$ induces a bijection on connected components*. Moreover, by 3.4.2, we have

$$\varphi_3^t f = (\Gamma\Xi)\varphi^t(p^2 f) = (\Gamma U)\Gamma(Tpf)\varphi^t(p^2 f) = (\Gamma U)\varphi_2^t(pf).$$

5.3.1. Consider now the following commutative square of pointed groupoids (Fig. 64)

$$
\begin{array}{ccc}
(\Gamma f)^t & \xrightarrow{(pf)^t} & x^t \\
\Big\downarrow{\scriptstyle \varphi^t f} & & \Big\downarrow{\scriptstyle \mathrm{Id}} \\
\Gamma(f^t) & \xrightarrow{p(f^t)} & x^t
\end{array}
$$

Fig. 64

Because the constructions on groupoids of paragraph 2 are "natural", it induces a commutative square (Fig. 65)

$$\begin{array}{ccc} \Omega(x^t) & \xrightarrow{\text{Id}} & \Omega(x^t) \\ \big\downarrow{\scriptstyle r(pf)^t} & & \big\downarrow{\scriptstyle rp(f^t)} \\ \Gamma p\,(pf)^t & \xrightarrow{\Gamma U} & \Gamma p^2(f^t) \end{array}$$

Fig. 65

which can be "composed" with the commutative square (Fig. 66)

$$\begin{array}{ccc} (\Omega x)^t & \xrightarrow{\varphi^t 0_x} & \Omega(x^t) \\ \big\downarrow{\scriptstyle (rpf)^t} & & \big\downarrow{\scriptstyle r(pf)^t} \\ (\Gamma p^2 f)^t & \xrightarrow{\varphi_2^t(pf)} & \Gamma p\,(pf)^t \end{array}$$

Fig. 66

obtained from the square (Sf) of 5.2 by replacing f by pf. If we consider then the equality $\varphi_3^t f = (\Gamma U)\,\varphi_2^t(pf)$ of 5.3, we see that we have

$$(rp\,(f^t))\,(\varphi^t 0_x) = (\varphi_3^t f)\,(rpf)^t.$$

Since we also have $(\varphi^t 0_x)\,(\sigma x)^t = \sigma(x^t)\,(\varphi^t 0_x)$ by 5.1, we then get a commutative square (Fig. 67)

$$\begin{array}{ccc} (\Omega x)^t & \xrightarrow{\varphi^t 0_x} & \Omega(x^t) \\ \big\downarrow{\scriptstyle (rpf)^t(\sigma x)^t} & & \big\downarrow{\scriptstyle (rp(f^t))\,\sigma(x^t)} \\ (\Gamma p^2 f)^t & \xrightarrow{\varphi_3^t f} & \Gamma p^2(f^t) \end{array}$$

Fig. 67

5.3.2. Consider finally the "cube" (Fig. 68)

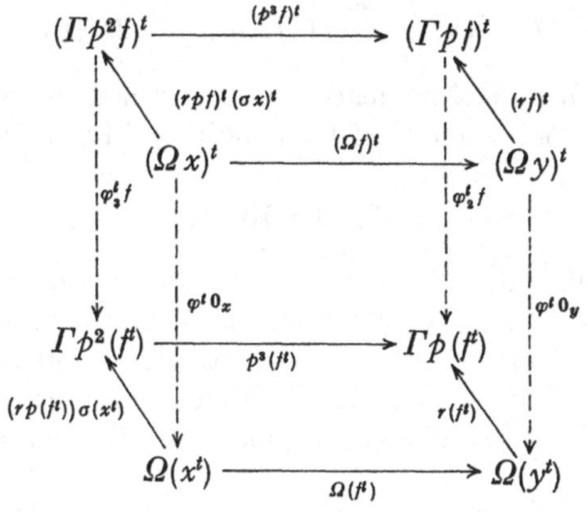

Fig. 68

The lateral faces this cube are commutative by 5.3.1, 5.3, 5.2, and 3.4.3. By 5.3, 5.2 and 3.2, the vertical arrows induce bijections on connected components. Since the base of our cube induces a commutative square of $\overline{\mathfrak{Gr}}$ (category of pointed groupoids modulo homotopy), the image of the top of the cube under Π_0 is commutative. Hence we have

$$|t, p^3 f| \circ |t, r p f| \circ |t, \sigma x| = |t, r f| \circ |t, \Omega f|.$$

5.3.3. Finally, we have to define the operation of Ωy on Γf i.e., we have to make the group $|t, \Omega y|$ operate on the set $|t, \Gamma f|$ functorially in t, and in a way compatible with the map $(qf)^t$. Since the square of Fig. 69

$$\begin{array}{ccc} (\Omega y)^t & \xrightarrow{(qf)^t} & (\Gamma f)^t \\ \downarrow{\scriptstyle \varphi_t 0_y} & & \downarrow{\scriptstyle \varphi_t f} \\ \Omega(y^t) & \xrightarrow{q(f^t)} & \Gamma(f^t) \end{array}$$

Fig. 69

is commutative, and since the maps $\Pi_0(\varphi_t 0_y)$ and $\Pi_0(\varphi_t f)$ are bijective, it is equivalent to make $\Pi_1(y^t)$ operate on $\Pi_0(\Gamma(f^t))$: hence we are brought back to the case of groupoids (see 2.4.1), and this concludes the proof.

In particular, if x is isomorphic in $\overline{\mathfrak{X}}$ to the zero object, it follows that $\Pi_0(\Omega y)^t$ *is isomorphic to* $\Pi_0(\Gamma f)^t$.

5.4. Note: It follows from Fig. 58 that, for each object t of $\mathfrak{X}$, the exact sequence of length 4

$$|t, \Omega x| \xrightarrow{|t, \Omega f|} |t, \Omega y| \xrightarrow{|t, qf|} |t, \Gamma f| \xrightarrow{|t, pf|} |t, x| \xrightarrow{|t, f|} |t, y|$$

is isomorphic to the sequence

$$\Pi_0 \Omega(x^t) \xrightarrow{\Pi_0 \Omega(f^t)} \Pi_0 \Omega(y^t) \xrightarrow{\Pi_0 q(f^t)} \Pi_0 \Gamma(f^t) \xrightarrow{\Pi_0 p(f^t)} \Pi_0(x^t) \xrightarrow{\Pi_0(f^t)} \Pi_0(y^t).$$

Hence, this *truncated* sequence can be defined for each 2-category satisfying conditions A and B of 1.4, condition C being unnecessary.

6. Duality

6.1. To put an end to this "long journey through the desert", we will show how the above results can be "dualized": for each 2-category $\mathfrak{Y}$, we write $\mathfrak{Y}^\circ$ for the *dual 2-category* of $\mathfrak{Y}$: this 2-category has the same objects as $\mathfrak{Y}$, but its "categories of morphisms" are defined by the equation $\mathcal{H}om_{\mathfrak{Y}^\circ}(x, y) = \mathcal{H}om_{\mathfrak{Y}}(y, x)$; moreover, the composition functor

$$\overset{\circ}{\mu}_{x,y,z}: \mathcal{H}om_{\mathfrak{Y}^\circ}(x, y) \times \mathcal{H}om_{\mathfrak{Y}^\circ}(y, z) \to \mathcal{H}om_{\mathfrak{Y}^\circ}(x, z)$$

of $\mathfrak{Y}^\circ$ is the composition of the canonical isomorphism from $\mathcal{H}om_{\mathfrak{Y}}(y,x)\times\mathcal{H}om_{\mathfrak{Y}}(z,y)$ onto $\mathcal{H}om_{\mathfrak{Y}}(z,y)\times\mathcal{H}om_{\mathfrak{Y}}(y,x)$, with the composition functor

$$\mu_{z,y,x}\colon \mathcal{H}om_{\mathfrak{Y}}(z,y)\times\mathcal{H}om_{\mathfrak{Y}}(y,x)\to\mathcal{H}om_{\mathfrak{Y}}(z,x).$$

We can say briefly that "we go from $\mathfrak{Y}^\circ$ to $\mathfrak{Y}$ by keeping the same object and the same 2-morphisms, while reversing the direction of the 1-morphisms". To each property of $\mathfrak{Y}^\circ$ corresponds a property of $\mathfrak{Y}$, called its dual. Let us give a few examples:

6.2. From now on, suppose that $\mathfrak{Y}$ satisfies conditions A and B of paragraph 3; then $\mathfrak{Y}^\circ$ also satisfies these conditions, which are self-dual. Condition C is "dualized" as follows:

Let $f\colon x\to y$ be a 1-morphism of $\mathfrak{Y}$ and let $\Psi f=(Cf, if, kf)$ be a triple formed by an object Cf, a 1-morphism $if\colon y\to Cf$, and a 2-morphism $kf\colon 0\to(if)f$. For each object t of $\mathfrak{Y}$, the triple Ψf determines a functor $\psi_t f\colon (Cf)_t\to\Gamma(f_t)$ (for each object z of $\mathfrak{Y}$, z_t denotes the groupoid $\mathcal{H}om_{\mathfrak{Y}}(z,t)$, and f_t the functor $\mathcal{H}om_{\mathfrak{Y}}(f,t)\colon y_t\to x_t\ldots$). More precisely, $\psi_t f$ associates with each 1-morphism $g\colon Cf\to t$ the pair $(g(if),\, g*(kf))$, which is an object of $\Gamma(f_t)$. Similarly, $\psi_t f$ associates with each 2-morphism $\gamma\colon g\to g'$ the 2-morphism $\gamma*(if)$, so that we have a commutative diagram (Fig. 70)

$$\begin{array}{ccccc}
(Cf)_t & \xrightarrow{(if)_t} & y_t & \xrightarrow{f_t} & x_t\\
\downarrow{\scriptstyle \psi_t f} & & \downarrow\wr & & \downarrow\wr\\
\Gamma(f_t) & \xrightarrow{p(f_t)} & y_t & \xrightarrow{f_t} & x_t
\end{array}$$

Fig. 70

We can say that the 2-category $\mathfrak{Y}^\circ$ satisfies condition C of 3.2 if and only if $\mathfrak{Y}$ satisfies condition C°:

C°. *For each 1-morphism $f\colon x\to y$ of $\mathfrak{Y}$, there is a triple (Cf, if, kf) such that $\psi_t f\colon (Cf)_t\to\Gamma(f_t)$ is connected for all objects t of $\mathfrak{Y}$.*

6.3. Let $\mathcal{A}r'\mathfrak{Y}$ be the following category: an object of $\mathcal{A}r'\mathfrak{Y}$ is a 1-morphism $f\colon x\to y$ of $\mathfrak{Y}$; the morphisms of $\mathcal{A}r'\mathfrak{Y}$ are the diagrams

$$(\varrho)\quad\begin{array}{ccc}
x & \xrightarrow{f} & y\\
{\scriptstyle u}\downarrow & \overset{\beta}{\curvearrowright} & \downarrow{\scriptstyle v}\\
x' & \xrightarrow[f']{} & y'
\end{array}$$

where u, v, f, f' are 1-morphisms and $\beta\colon f'u\to vf$ a 2-morphism. The domain (resp. range) of ϱ is f (resp. f'). We define the composition of

$$(\varrho)\quad\begin{array}{ccc}
x & \xrightarrow{f} & y\\
{\scriptstyle u}\downarrow & \overset{\beta}{\curvearrowright} & \downarrow{\scriptstyle v}\\
x' & \xrightarrow[f']{} & y'
\end{array}\qquad\text{with}\qquad(\varrho')\quad\begin{array}{ccc}
x' & \xrightarrow{f'} & y'\\
{\scriptstyle u'}\downarrow & \overset{\beta'}{\curvearrowright} & \downarrow{\scriptstyle v'}\\
x'' & \xrightarrow[f'']{} & y''
\end{array}$$

as the diagram

$$(\varrho'') \qquad \begin{array}{ccc} x & \xrightarrow{f} & y \\ {\scriptstyle u'u}\big\downarrow & \overset{\beta''}{\curvearrowright} & \big\downarrow{\scriptstyle v'v} \\ x'' & \xrightarrow[f'']{} & y'' \end{array}$$

where $\beta'' = (v' * \beta) \circ (\beta' * u)$.

6.3.1. From now on, suppose that $\mathfrak{Y}$ satisfies A, B and C°, and let us associate with each object $f\colon x \to y$ of $\mathscr{A}r'\mathfrak{Y}$ a triple (Cf, if, kf) satisfying C°. Let us associate also with each morphism

$$(\varrho) \qquad \begin{array}{ccc} x & \xrightarrow{f} & y \\ {\scriptstyle u}\big\downarrow & \overset{\beta}{\curvearrowright} & \big\downarrow{\scriptstyle v} \\ x' & \xrightarrow[f']{} & y \end{array}$$

of $\mathscr{A}r'\mathfrak{Y}$ a 1-morphism $C\varrho\colon Cf \to Cf'$ such that $(C\varrho)(if) = (if')v$ and $(C\varrho) * (kf) = ((if') * \beta) \circ ((kf') * u)$. When y is the zero object of $\mathfrak{Y}$, we write Σx instead of Cf, and we say that Σx is the *suspension* of x. The functor $t \rightsquigarrow |\Sigma x, t|$ is isomorphic to the functor which associates with t the Poincaré group of the groupoid $\mathscr{H}om_{\mathfrak{Y}}(x, t)$ at the point 0_t^x. *Hence the suspension of x is given a natural $\mathfrak{Y}$-cogroup structure.*

When ϱ is the morphism

$$\begin{array}{ccc} 0 & \longrightarrow & x \\ \big\downarrow & \overset{\mathrm{Id}}{\curvearrowright} & \big\downarrow{\scriptstyle f} \\ 0 & \longrightarrow & y \end{array}$$

we write Σf instead of $C\varrho$. Thus we obtain a functor Σ from $\mathfrak{Y}_{0,1}$ (1.3) to $\mathfrak{Y}$; this functor induces, by passing to the quotient, a functor (still written Σ) from $\mathfrak{Y}$ to $\mathfrak{Y}$.

When ϱ is the morphism

$$\begin{array}{ccc} x & \xrightarrow{f} & y \\ {\scriptstyle \mathrm{Id}\,x}\big\downarrow & \overset{\mathrm{Id}}{\curvearrowright} & \big\downarrow \\ x & \longrightarrow & 0 \end{array}$$

we write jf instead of $C\varrho$.

6.3.2. The maps which associate with each object $f\colon x \to y$ of $\mathscr{A}r'\mathfrak{Y}$ the sequence

$$(*) \qquad x \xrightarrow{[f]} y \xrightarrow{[if]} Cf \xrightarrow{[jf]} \Sigma x \xrightarrow{[\Sigma f]} \Sigma y \xrightarrow{[\Sigma if]} \Sigma Cf \xrightarrow{[\Sigma jf]} \Sigma^2 x \ldots$$

of $\mathfrak{Y}$, and with each morphism

$$(\varrho)\qquad \begin{array}{ccc} x & \xrightarrow{f} & y \\ {\scriptstyle u}\downarrow & \overset{\beta}{\curvearrowright} & \downarrow{\scriptstyle v} \\ x & \xrightarrow[f']{} & y' \end{array}$$

of $\mathscr{A}r'\mathfrak{Y}$ the morphism of Fig. 71

$$\begin{array}{ccccccccccc} x & \xrightarrow{[f]} & y & \xrightarrow{[if]} & Cf & \xrightarrow{[jf]} & \Sigma x & \xrightarrow{[\Sigma f]} & \Sigma y & \xrightarrow{[\Sigma if]} & \Sigma Cf \\ {\scriptstyle [u]}\downarrow & & {\scriptstyle [v]}\downarrow & & {\scriptstyle [C\varrho]}\downarrow & & {\scriptstyle [\Sigma u]}\downarrow & & {\scriptstyle [\Sigma v]}\downarrow & & {\scriptstyle [\Sigma C\varrho]}\downarrow \\ x' & \xrightarrow{[f']} & y' & \xrightarrow{[if']} & Cf' & \xrightarrow{[jf']} & \Sigma x' & \xrightarrow{[\Sigma f']} & \Sigma y' & \xrightarrow{[\Sigma if']} & \Sigma Cf' \end{array}$$

Fig. 71

determine a functor from $\mathscr{A}r'\mathfrak{Y}$ to the category of sequences of $\mathfrak{Y}$. The objects $\Sigma x, \Sigma y, \Sigma Cf, \ldots$ are cogroups of $\mathfrak{Y}$, and the morphisms $[\Sigma f]$, $[\Sigma if] \ldots$ are cogroup homomorphisms.

If different values are chosen for Cf, if, kf, $C\varrho$, the above functor will be replaced by an isomorphic functor.

Finally, the sequence $(*)$ is l-exact. In other words, for each object t of $\mathfrak{Y}$, the sequence of pointed sets

$$\ldots |\Sigma y, t| \xrightarrow{|\Sigma f, t|} |\Sigma x, t| \xrightarrow{|jf, t|} |Cf, t| \xrightarrow{|if, t|} |y, t| \xrightarrow{|f, t|} |x, t|$$

is exact.

6.4. Suppose now that $\mathfrak{X} = \mathfrak{Y}$ satisfies conditions A, B, C and C°. It follows then from 6.3.1 and 3.3 that the sets $|\Sigma x, y|$ and $|x, \Omega y|$ are both identified with the Poincaré group of $\mathscr{H}om_{\mathfrak{Y}}(x, y)$ at the point 0^x_y. *Consequently, the functor* $x \rightsquigarrow \Sigma x$ *from* $\overline{\mathfrak{X}}$ *to* $\overline{\mathfrak{X}}$ *is left adjoint to the functor* $y \rightsquigarrow \Omega y$.

Under the same conditions, if y is a group of $\overline{\mathfrak{X}}$, it follows from a well-known argument that the two group structures on $|\Sigma x, y| = |x, \Omega y|$, defined respectively by Σx and y, coincide, and make $|x, \Omega y|$ an abelian group. Hence Ωy is *an abelian group of* $\overline{\mathfrak{X}}$; in particular, $\Omega^2 t$ *is always an abelian group of* $\overline{\mathfrak{X}}$; *similarly,* $\Sigma^2 t$ *is an abelian cogroup of* $\overline{\mathfrak{X}}$.

Still under the same conditions, consider the following morphism (ϱ) of $\mathscr{A}r'\mathfrak{Y}$

$$(\varrho)\qquad \begin{array}{ccc} \Gamma f & \longrightarrow & o \\ {\scriptstyle pf}\downarrow & \overset{(hf)^{-1}}{\curvearrowright} & \downarrow \\ x & \xrightarrow[f]{} & y \end{array}$$

[recall that hf is a morphism of the groupoid $\mathscr{H}om_{\mathfrak{Y}}(\Gamma f, y)$, and hence that it is invertible]. Then $C(\varrho)$ is a morphism $\Sigma\Gamma f \to Cf$ of $\mathfrak{Y}$. We could define in a similar way a canonical morphism $\Gamma f \to \Omega Cf$.

7. First Example: Pointed Topological Spaces

7.1. Let $\boldsymbol{X}$ and $\boldsymbol{Y}$ be two topological spaces with base points $\boldsymbol{x}_0$ and $\boldsymbol{y}_0$ respectively, and $\mathcal{H}om.(\boldsymbol{X}, \boldsymbol{Y})$ the pointed complex whose n-th component is the set of continuous maps h from $\Delta^n \times \boldsymbol{X}$ to $\boldsymbol{Y}$, which send $\Delta^n \times \boldsymbol{x}_0$ to $\boldsymbol{y}_0$ (Δ^n is the geometric simplex of dimension n). With such a map h, we can associate the map $\tilde{h}$: $(d, x) \rightsquigarrow (d, h(d, x))$ from $\Delta^n \times \boldsymbol{X}$ to $\Delta^n \times \boldsymbol{Y}$. If $\boldsymbol{Z}$ is a third complex, with base point $\boldsymbol{z}_0$, we can then define a morphism of complexes

$$\nu_{\boldsymbol{X},\boldsymbol{Y},\boldsymbol{Z}}: \mathcal{H}om.(\boldsymbol{X}, \boldsymbol{Y}) \times \mathcal{H}om.(\boldsymbol{Y}, \boldsymbol{Z}) \to \mathcal{H}om.(\boldsymbol{X}, \boldsymbol{Z})$$

by sending the pair (h, k) belonging to $\mathcal{H}om.(\boldsymbol{X}, \boldsymbol{Y})_n \times \mathcal{H}om.(\boldsymbol{Y}, \boldsymbol{Z})_n$ to the map l: $\Delta^n \times \boldsymbol{X} \to \boldsymbol{Z}$ defined by the equation $\tilde{l} = \tilde{k} \circ \tilde{h}$.

Since the functor Π, which associates with each complex K its Poincaré groupoid ΠK, commutes with finite products (II, 7.5), $\nu_{\boldsymbol{X},\boldsymbol{Y},\boldsymbol{Z}}$ induces a functor

$$\mu_{\boldsymbol{X},\boldsymbol{Y},\boldsymbol{Z}}: \Pi\, \mathcal{H}om.(\boldsymbol{X}, \boldsymbol{Y}) \times \Pi\, \mathcal{H}om.(\boldsymbol{Y}, \boldsymbol{Z}) \to \Pi\, \mathcal{H}om.(\boldsymbol{X}, \boldsymbol{Z}).$$

It is then clear that we can define a 2-category $.\mathfrak{Top}$ as follows: the objects are the pointed topological spaces, the categories $\mathcal{H}om_{.\mathfrak{Top}}(\boldsymbol{X}, \boldsymbol{Y})$ are the groupoids $\Pi\, \mathcal{H}om.(\boldsymbol{X}, \boldsymbol{Y})$, and the composition functors are the functors $\mu_{\boldsymbol{X},\boldsymbol{Y},\boldsymbol{Z}}$ induced by $\nu_{\boldsymbol{X},\boldsymbol{Y},\boldsymbol{Z}}$. The 1-morphisms of $.\mathfrak{Top}$ are the continuous maps which send base points to base points; similarly, we see easily that we can describe the 2-morphisms as follows: let α and β be two maps from $\boldsymbol{I} \times \boldsymbol{X}$ to $\boldsymbol{Y}$ such that $\alpha(t, \boldsymbol{x}_0) = \beta(t, \boldsymbol{x}_0) = \boldsymbol{y}_0$ $(\boldsymbol{I} = [0, 1])$; we will say that α and β are homotopic if there is a map $\boldsymbol{h}$: $\boldsymbol{I} \times \boldsymbol{I} \times \boldsymbol{X} \to \boldsymbol{Y}$ such that $\boldsymbol{h}(s, t, \boldsymbol{x}_0) = \boldsymbol{y}_0$, $\boldsymbol{h}(0, t, \boldsymbol{x}) = \alpha(t, \boldsymbol{x})$, $\boldsymbol{h}(1, t, \boldsymbol{x}) = \beta(t, \boldsymbol{x})$, $\boldsymbol{h}(s, 0, \boldsymbol{x}) = \alpha(0, \boldsymbol{x}) = \beta(0, \boldsymbol{x})$ and $\boldsymbol{h}(s, 1, \boldsymbol{x}) = \alpha(1, \boldsymbol{x}) = \beta(1, \boldsymbol{x})$; this relation is obviously an equivalence relation, and we write $[\alpha]$ for the equivalence class of α. If $\boldsymbol{f}$ and $\boldsymbol{g}$ denote the maps $\boldsymbol{x} \rightsquigarrow \alpha(0, \boldsymbol{x})$ and $\boldsymbol{x} \rightsquigarrow \alpha(1, \boldsymbol{x})$, $\boldsymbol{f}$ and $\boldsymbol{g}$ depend only on $[\alpha]$, so that $[\alpha]$ is an equivalence class of homotopies between $\boldsymbol{f}$ and $\boldsymbol{g}$. With the notations of 1.3, we then have $\mathfrak{d}_1[\alpha] = \boldsymbol{f}$ and $\mathfrak{r}_1[\alpha] = \boldsymbol{g}$; the composition laws given explicitely in 1.3 are the "obvious" laws.

7.2. Let us show now that the 2-category $.\mathfrak{Top}$ satisfies conditions A, B and C of paragraph 3: condition A follows from the definition; for o, it is sufficient to take a topological space consisting of a single point. Then the objects of the category $.\overline{\mathfrak{Top}}$ are the pointed topological spaces, its morphisms are the homotopy classes of continuous maps which send base points to base points.

If $\boldsymbol{f}$: $\boldsymbol{X} \to \boldsymbol{Y}$ is a continuous map which respects the base points $\boldsymbol{x}_0$ and $\boldsymbol{y}_0$, we construct $\Gamma \boldsymbol{f}$, $p\boldsymbol{f}$ and $h\boldsymbol{f}$ as follows: first, let $\boldsymbol{E}(\boldsymbol{Y}, \boldsymbol{y}_0)$ be the function space whose points are the continuous maps γ: $\boldsymbol{I} \to \boldsymbol{Y}$ such that

$\gamma(0)=y_0$, its topology being the compact open topology; then Γf is the subspace of $E(Y, y_0)\times X$ formed by all pairs (γ, x) such that $\gamma(1)=f(x)$; similarly, pf sends (γ, x) to x, and we define $(hf)(t, \gamma, x)=\gamma(t)$.

7.3. The 2-category of pointed topological spaces also satisfies condition C° of 6.2: for each map $f\colon X\to Y$ which sends the base point x_0 of X to the base point y_0 of Y, we can construct the triple (Cf, if, kf) of 6.1 as follows: if Z and T are two pointed topological spaces with base points z_0 and t_0 respectively, write $Z\wedge T$ for the pointed space obtained from $Z\times T$ by identifying the points of $Z\times\{t_0\}\cup\{z_0\}\times T$; the image of this subset in $Z\wedge T$ will then consist of a single point, which will be the base point.

In particular, if 0 is taken as a base point for the segment $I=[0, 1]$, $I\wedge X$ is obtained from the cone with base X by identifying the points of the generating line through x_0. The canonical isomorphism uf from X onto the base of $I\wedge X$ is thus obtained by the composition of the map $x\rightsquigarrow(1, x)$ from X to $I\times X$ with the canonical projection of $I\times X$ onto $I\wedge X$. For Cf, we choose the amalgamated sum of the diagram

$$\begin{array}{ccc} X & \xrightarrow{uf} & I\wedge X \\ {\scriptstyle f}\downarrow & & \\ Y & & \end{array}$$

the base point of Cf being the image of y_0 under the canonical injection if of Y into the amalgamated sum. Finally, the 2-morphism kf must be an equivalence class of homotopies between the zero map from X to Cf and the map $(if)\circ f$; for kf, we take the equivalence class of the composition

$$I\times X\xrightarrow{\text{can. proj.}} I\wedge X\xrightarrow{\text{can. map}} Cf.$$

7.4. Suppose now that, with the notation of 7.3, f is the injection of a subspace X of Y into Y. We say that the pair (Y, X) has *the homotopy extension property* if each continuous map from $(I\times X)\cup\{0\}\times Y$ to a space T can be extended to $I\times Y$.

When L is a simplicial set and K is a subcomplex of L, the pair $(|L|, |K|)$ satisfies the homotopy extension property. Indeed, by chapter III, if i denotes the inclusion of $(\Delta[1]\times K)\cup(\{0\}\times L)$ into $\Delta[1]\times L$, $|i|$ is identified with the inclusion of $(I\times|K|)\cup(\{0\}\times|L|)$ into $I\times|L|$. Since i is an anodyne extension of $\Delta^\circ\mathcal{E}$, $(I\times|K|)\cup(\{0\}\times|L|)$ is a deformation retract of $I\times|L|$: our statement follows immediately from this fact (see VII, 1.7).

7.5. *Lemma: Let* (Y, X) *be a pair with the homotopy extension property, and let* $h\colon I\times X\to X$ *be a retracting deformation of* X *onto* x_0. *Then the canonical projection of* Y *onto* Y/X *is a homotopy equivalence.*

Recall first that a retracting deformation of X onto x_0 is a map h such that $h(1, x)=h(t, x_0)=x_0$ and $h(0, x)=x$. On the other hand,

Y/X denotes the quotient obtained from Y by identifying the points of X, the image of X in Y/X being the base point.

Let g be the map from $(I\times X)\cup(\{0\}\times Y)$ to Y whose restriction to $I\times X$, is h, the restriction to $\{0\}\times Y$ being the identity map of Y. On the other hand, let f be an extension of g to $I\times Y$. Then the map f_1: $y\rightsquigarrow f(1, y)$ is the composition of the canonical projection $p: Y\to Y/X$ with a map $s: Y/X\to Y$. By definition of s, $s\circ p$ is homotopic to Id Y. On the other hand, f defines, by taking the quotient, a map $e: I\times(Y/X)\to Y/X$ such that $e(0, z)=z$ and $e(1, z)=p\circ s$.

7.6. If (Y, X) has the homotopy extension property (f being the indusion of X into Y), the pair $(Cf, I\wedge X)$ also has this property: let $g: I\times(I\wedge X)\cup\{0\}\times Cf\to T$ be a continuous map. Since $I\times Cf$ is simply the amalgamated sum of the diagram

$$\begin{array}{ccc} I\times X & \xrightarrow{I\times uf} & I\times(I\wedge X) \\ \downarrow{\scriptstyle I\times f} & & \\ I\times Y & & \end{array}$$

an extension of g to $I\times Cf$ is uniquely defined by an extension to $I\times Y$ of the restriction of g to $I\times X\cup\{0\}\times Y$.

In particular, transitivity of amalgamated sums imply the existence of a canonical isomorphism from $Cf/I\wedge X$ onto Y/X. Hence we have a projection p of Cf onto Y/X. Since $I\wedge X$ is contractible, lemma 7.5 implies the following proposition.

Proposition: If the pair (Y, X) has the homotopy extension property, f denoting the inclusion of X into Y, the canonical projection p of Cf onto Y/X is a homotopy equivalence.

If we replace Cf by Y/X in PUPPE's exact sequence

$$\ldots [\Sigma Cf, T]\to[\Sigma Y, T]\to[\Sigma X, T]\to[Cf, T]\to[Y, T]\to[X, T]$$

we get the exact sequence

$$\ldots [\Sigma(Y/X), T]\to[\Sigma Y, T]\to[\Sigma X, T]\to[Y/X, T]\to[Y, T]\to[X, T]$$

called BARRATT'S exact sequence.

[According to general use, we used the notation $[X, T]$ instead of our notations $|X, T|$ or $\mathfrak{Top}\,(X, T)$.]

8. Second Example: Differential Complexes of an Abelian Category

8.1. If $\mathscr{A}$ is an abelian category, a differential complex, (or simply a complex) of $\mathscr{A}$ is a sequence X of pairs (X_n, d_n^X) where X_n is an object of $\mathscr{A}$ and d_n^X a morphism $d_n^X: X_{n+1}\to X_n$, such that $d_{n-1}^X\circ d_n^X=0$. Usually, we write d_n or simply d instead of d_n^X, and we say that d is the boundary operator, or the differential. If X and Y are two complexes, a morphism $f\colon X\to Y$ is a sequence of morphisms $f_n\colon X_n\to Y_n$ of $\mathcal{A}$ such that $d_n\circ f_{n+1}=f_n\circ d_n$ for all $n\in\mathbb{Z}$. These morphisms are composed in an obvious way, and determine a category $K(\mathcal{A})$ called *the category of complexes of* $\mathcal{A}$.

8.2. If $\mathcal{A}$ is the category $\mathcal{Ab}$ of abelian groups, we will first associate with each complex X of abelian groups a groupoid ΠX: the set of objects of ΠX is $\ker d_{-1}$; the set of morphisms of ΠX is the direct sum $\ker d_{-1}\oplus \operatorname{coker} d_1$; since $d_0\circ d_1$ is zero, d_0 induces a map d from $\operatorname{Coker} d_1$ to $\operatorname{Ker} d_{-1}$; the domain of a morphism (x,f) of ΠX is then chosen to be x, while the range is $x+df$; finally, composition is defined by the formula:

$$(x+df,g)\,(x,f)=(x,f+g).$$

Now, let X and Y be two complexes of abelian groups. Recall that the tensor product $X\otimes Y$ is the complex Z defined by the following conditions: the n-th component Z_n of Z is the direct sum of the abelian group $X_p\otimes Y_q$, the sum being taken over all pairs (p,q) such that $p+q=n$; if x belongs to X_p and y to Y_q, we have definition

$$d(x\otimes y)=(dx)\otimes y+(-1)^p x\otimes (dy)$$

the first term on the right side belonging to the summand $X_{p-1}\otimes Y_q$ of Z_{n-1}, the other to the summand $X_p\otimes Y_{q-1}$.

In the sequel, we will write C for the canonical functor from $\Pi X\times\Pi Y$ to $\Pi(X\otimes Y)$ defined as follows: if (x,y) is an object of $\Pi X\times\Pi Y$, i.e. an element of $X_0\times Y_0$, $C(x,y)$ is simply $x\otimes y$; if $((x,f),(y,g))$ is a morphism of $\Pi X\times\Pi Y$ and if f' and g' denote representatives of f and g in X_1 and Y_1, $C((x,f),(y,g))$ is the pair $(x\otimes y,h)$ where h denotes the canonical image $(x+df)\otimes g+f\otimes y$ of $(x+df)\otimes g'+f'\otimes y$ in Z_1/d_1Z_2.

8.3. Let us return now to the general case, let X and Y be two complexes of an abelian category $\mathcal{A}$, and let $\mathcal{Hom}(X,Y)$ be the following complex of abeling groups: its n-th component is the product $\prod_{p\in Z}\mathcal{A}(X_p,Y_{p+n})$. If $f=(f_p)_{p\in\mathbb{Z}}$ is an element of this product, df is defined by the equality

$$d(f_p)_{p\in\mathbb{Z}}=(df_p-(-1)^n f_{p-1}d)_{p\in\mathbb{Z}}.$$

If X, Y and Z are three complexes of $\mathcal{A}$, we write

$$\nu_{X,Y,Z}\colon \mathcal{Hom}(X,Y)\otimes\mathcal{Hom}(Y,Z)\to\mathcal{Hom}(X,Z)$$

for the morphism of complexes which sends the element $(f_p)\otimes(g_q)$ of $\mathcal{Hom}_m(X,Y)\otimes\mathcal{Hom}_n(Y,Z)$ to the element $((-1)^{mn}g_{p+m}\circ f_p)$ of $\mathcal{Hom}_{m+n}(X,Z)$.

8.4. We are now ready to define a new 2-category $\mathfrak{K}(\mathscr{A})$, which we will call the 2-category of complexes of $\mathscr{A}$: an object of $\mathfrak{K}(\mathscr{A})$ is a differential complex of $\mathscr{A}$; if X and Y are two such complexes, the category of morphisms $\mathscr{H}om_{\mathfrak{K}(\mathscr{A})}(X, Y)$ is the groupoid $\Pi\,\mathscr{H}om(X, Y)$ defined by the complex $\mathscr{H}om(X, Y)$ (8.2). Finally, the composition functors

$$\mu_{X,Y,Z}\colon \Pi\,\mathscr{H}om(X, Y)\times\Pi\,\mathscr{H}om(Y, Z)\to\Pi\,\mathscr{H}om(X, Z)$$

are obtained by the composition of the canonical functor from

$$\Pi\,\mathscr{H}om(X, Y)\times\Pi\,\mathscr{H}om(Y, Z) \quad\text{to}\quad \Pi(\mathscr{H}om(X, Y)\otimes\mathscr{H}om(Y, Z))$$

defined in 8.2, with the functor $\Pi\nu_{X,Y,Z}$ induced by $\nu_{X,Y,Z}$ (8.3).

The 1-morphisms of domain X and range Y are the morphisms of complexes from X to Y. If f and g are two such 1-morphisms, a 2-morphism $f\to g$ is a pair (f, τ) whose last component is a class of elements t of $\mathscr{H}om_1(X, Y)$ such that $g-f=dt+td$; two elements t and s of $\mathscr{H}om_1(X, Y)$ belong to the same class of they are homotopic, i.e. if there is a $u\in\mathscr{H}om_2(X, Y)$ such that $t-s=du-ud$.

8.5. The 2-category of complexes of $\mathscr{A}$ satisfies conditions A, B and C of 3.1. Condition A is clear, as well as the existence of a zero object (it is the complex with all components equal to zero). Take then a 1-morphism $f\colon X\to Y$; we can then construct the triple $(\Gamma f, pf, hf)$ as follows:

Γf is the "simple total" complex associated with the "double" complex of Fig. 72

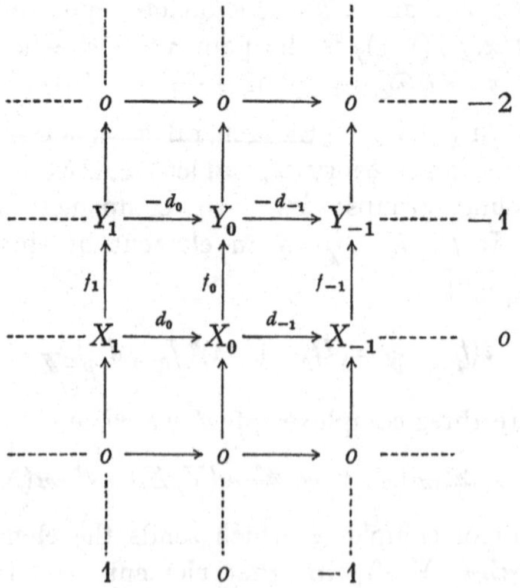

Fig. 72

We then have $(\Gamma f)_n = X_n \oplus Y_{n+1}$, the boundary operator $d\colon (\Gamma f)_n \to (\Gamma f)_{n-1}$ inducing on X_n the morphism $X_n \to X_{n-1} \oplus Y_n$ with components d_{n-1} and f_n, and on Y_{n+1} the morphism $Y_{n+1} \to X_{n-1} \oplus Y_n$ with components 0 and $-d$. The components of the 1-morphism $pf\colon \Gamma f \to X$ are the canonical projections of $X_n \oplus Y_{n+1}$ onto X_n $(n \in \mathbb{Z})$. Finally, we choose hf to be the class of the element t of $\mathcal{H}om_1(\Gamma f, Y)$ whose n-th component is the canonical projection of $(\Gamma f)_n = X_n \oplus Y_{n+1}$ onto Y_{n+1}.

Note that with the choice we made for Γf, the space of loops ΩY of a complex Y is defined by the formulas:

$$(\Omega Y)_n = Y_{n+1} \quad \text{and} \quad d_n^{\Omega Y} = -d_{n+1}^Y.$$

8.6. The 2-category of complexes of $\mathcal{A}$ also satisfies condition C°, which is dual to C (6.2). For each 1-morphism $f\colon X \to Y$, we can construct the triple (Cf, if, kf) as follows:

Cf is the "simple total" complex associated with the "double" complex of Fig. 73

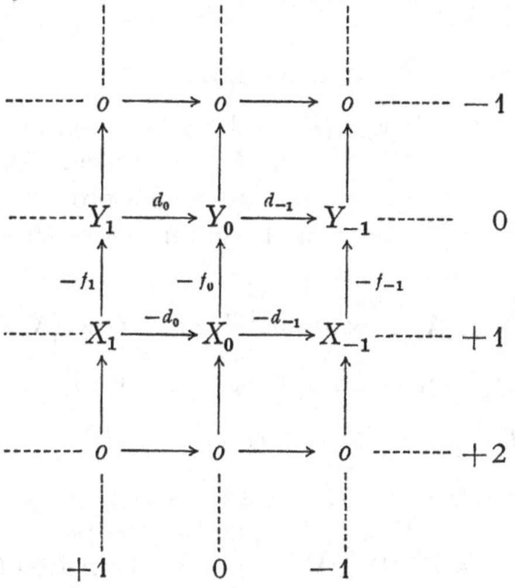

Fig. 73

Hence we have $(Cf)_n = X_{n-1} \oplus Y_n$, the boundary operator inducing on X_{n-1} a morphism with components $-d_{n-2}$ and $-f_{n-1}$, and on Y_n a morphism with components 0 and d_{n-1}. The n-th component of the 1-morphism $if\colon Y \to Cf$ is the canonical monomorphism from Y_n into the direct sum $X_{n-1} \oplus Y_n$. Finally, we choose kf to be the class of the elements s of $\mathcal{H}om_1(X, Cf)$ whose n-th component is the additive inverse of the canonical monomorphism from X_n into $(Cf)_{n+1} = X_n \oplus Y_{n+1}$.

Note that with this choice of Cf, the suspension ΣX is defined by the formulas: $(\Sigma X)_n = X_{n-1}$ and $d_n^{\Sigma X} = -d_{n-1}^X$. Hence we have $\Sigma = \Omega^{-1}$ and $Cf = \Sigma(\Gamma f)$ (see 6.4).

Remark: Let H_n be the homology functor from $K(\mathscr{A})$ to $\mathscr{A}$, i.e. the functor $X \rightsquigarrow \ker d^X_{n-1}/\operatorname{coker} d^X_n$ $(n \in \mathbb{Z})$. We can see that with each r-exact sequence of complexes $X \xrightarrow{f} Y \xrightarrow{g} Z$, the functor H_n associates an *exact* sequence

$$H_n(X) \xrightarrow{H_n(f)} H_n(Y) \xrightarrow{H_n(g)} H_n(Z).$$

If $f: X \to Y$ is an epimorphism, $\ker f$ and Γf have the same homology for all n; if f is a monomorphism, $\operatorname{coker} f$ and Cf have same homology for all n.

Chapter Six

Exact Sequences of the Homotopic Category

We will give here some applications of the preceding chapter to simplicial sets. We will also seize the opportunity to prove statements which will be used in the proof of the fundamental theorem of chapter VII.

1. Spaces of Loops

1.1. Let (X, x_0), (Y, y_0), (Z, z_0) be three pointed simplicial sets, and let us write $\Pi^{\cdot}(X, Y)$ for the Poincaré groupoid of the complex $\mathscr{H}om.(X, Y)$. Since the Poincaré groupoid of a product of simplicial sets is identified with the product of the groupoids of these simplicial sets (II, 7.5), the morphism

$$\mathscr{H}om.(X, Y) \times \mathscr{H}om.(Y, Z) \to \mathscr{H}om.(X, Z)$$

defined in IV, 4.2, induces, by passing to Poincaré groupoids, a functor

$$\Pi^{\cdot}_{X,Y,Z}: \Pi^{\cdot}(X, Y) \times \Pi^{\cdot}(Y, Z) \to \Pi^{\cdot}(X, Z).$$

Besides, it is clear that the $\Pi^{\cdot}_{X,Y,Z}$ are the composition functors of a 2-category whose objects are the pointed complexes: this 2-category satisfies conditions A and B of V, 1, and will be written $((.\Delta^{\circ}\mathscr{E}))$.

Since the set of connected components of a simplicial set is simply the set of connected components of its Poincaré groupoid, we see that $.\overline{\Delta^{\circ}\mathscr{E}}(X, Y) = \Pi^{\cdot}_0(X, Y)$ is simply $|X, Y|$ (see IV, 4.2 and V, 1.6). Hence there is no reason to make a distinction between the categories $.\overline{\Delta^{\circ}\mathscr{E}}$ and $((.\overline{\Delta^{\circ}\mathscr{E}}))$ defined in IV, 4.2 and V, 1.6. We will take care, however, not to identify the notations $|X, Y|$ and $.\mathscr{H}(X, Y)$ when Y is not a Kan complex (IV, 4.3).

1.2. Let $.\mathfrak{Kan}$ be the full sub-2-category of $((.\Delta^{\circ}\mathscr{E}))$ whose objects are the pointed Kan complexes: this sub-2-category has then less objects than $((.\Delta^{\circ}\mathscr{E}))$, but its categories of morphisms and its composition functors are $\Pi^{\cdot}(X, Y)$ and $\Pi^{\cdot}_{X,Y,Z}$ when X, Y and Z are Kan complexes.

The 2-category $.\mathfrak{Kan}$, which obviously satisfies A and B, also satisfies condition C of V, 3:

Let X and Y be two pointed Kan complexes, x_0 and y_0 their base points, and $f\colon X \to Y$ a morphism such that $f(x_0) = y_0$; we can then construct Γf, pf and hf as follows:

For Γf, take the fibred product of the diagram

$$\begin{array}{ccc} & & \mathcal{H}om.(\Delta[1], Y) \\ & & \downarrow e \\ X & \xrightarrow{f} & Y \end{array}$$

where e is induced by the inclusion of $\dot{\Delta}[1]$ into $\Delta[1]$ when Y is identified with $\mathcal{H}om.(\dot{\Delta}[1], Y)$ (IV, 4.1.2). It follows from the lemma below and from IV, 3.1.1 (ii bis) that Γf is also a Kan complex.

For pf, we take the canonical projection pr_1 of the fibred product onto the factor X. As for the canonical projection pr_2 of Γf onto $\mathcal{H}om.(\Delta[1], Y)$, it is canonically associated, by IV, 4.1.2, to a morphism $\Delta[1] \wedge \Gamma f \to Y$, i.e. to a homotopy connecting the zero morphism with $e \circ \mathrm{pr}_2 = f \circ \mathrm{pr}_1$. The equivalence class of this homotopy, in other words its canonical image in $\Pi^{\cdot}(\Gamma f, Y)$, will be precisely hf.

1.2.1. *Lemma: Let Y be a pointed Kan complex and $i\colon K \to L$ a monomorphism of pointed complexes. Then the morphism*

$$\mathcal{H}om.(i, Y)\colon \mathcal{H}om.(L, Y) \to \mathcal{H}om.(K, Y)$$

is a fibration.

The lemma follows form IV, 3.1.3, IV, 3.1.1 (ii bis) and from the fact that the square of Fig. 74

$$\begin{array}{ccc} \mathcal{H}om.(L, Y) & \xrightarrow{\text{incl.}} & \mathcal{H}om(L, Y) \\ {\scriptstyle \mathcal{H}om.(i, Y)}\downarrow & & \downarrow{\scriptstyle \mathcal{H}om(i, Y)} \\ \mathcal{H}om.(K, Y) & \longrightarrow & \mathcal{H}om(K, Y) \end{array}$$

Fig. 74

is cartesian.

1.2.2. It remains to be seen that the triple $(\Gamma f, pf, hf)$ which we have just constructed satisfies condition C of V, 3: let $t\colon T \to X$ be a 1-morphism of $.\mathfrak{Kan}$ and $h\colon 0 \to f \circ t$ a 2-morphism; we must look first for a 1-morphism $\tau\colon T \to \Gamma f$ such that $t = (pf) \circ \tau$ and $h = (hf) * \tau$. In order to do this, consider a homotopy $H\colon \Delta[1] \wedge T \to Y$ whose equivalence class is h (see IV, 5.3); the restriction of H to $T = \dot{\Delta}[1] \wedge T$ is $f \circ t$, so that the morphism $H_1\colon T \to \mathcal{H}om.(\Delta[1], Y)$, canonically associated with H, satisfies the equality $e \circ H_1 = f \circ t$. Hence we can choose for τ the morphism with components t and H_1.

Consider now two 1-morphisms $\tau, \tau'\colon T \rightrightarrows \Gamma f$ and let $t = (pf) \circ \tau$, $t' = (pf) \circ \tau'$, $h = (hf) * \tau$ and $h' = (hf) * \tau'$. We still have to show that, for each 2-morphism $a\colon t \to t'$ such that $(f * a) \circ h = h'$, there is a 2-morphism $\alpha\colon \tau \to \tau'$ such that $a = (pf) * \alpha$. In order to do this, let us examine the product $\Delta[1] \times \Delta[1] \times T$ (Fig. 75)

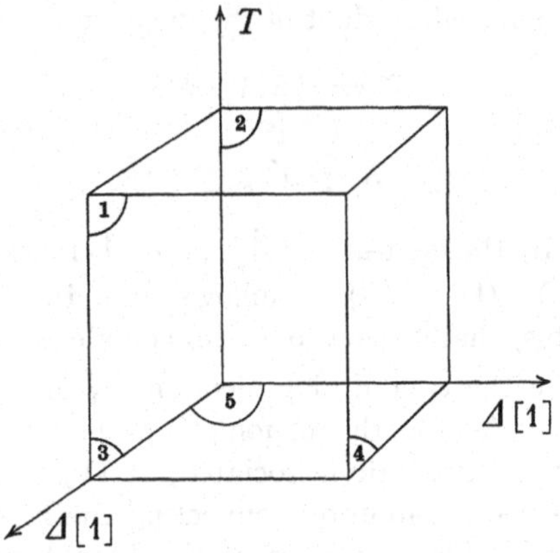

Fig. 75

Let $H\colon \Delta[1] \times \{0\} \times T \to Y$ be a morphism such that the equivalence class of the composition of H with the canonical isomorphism from $\Delta[1] \times T$ onto $\Delta[1] \times \{0\} \times T$ is h (H is thus defined on face 3 of our Figure). Similarly, let $H'\colon \Delta[1] \times \{1\} \times T \to Y$ be a morphism corresponding to h' (H' is defined on face 4). Finally, let A be a homotopy with equivalence class a, and $B\colon \{1\} \times \Delta[1] \times T \to Y$ the morphism induced by $f \circ A$ (B is defined on face 1). Since we have $h' = (f * a) \circ h$, there is a morphism $\alpha_1\colon \Delta[1] \times \Delta[1] \times T \to Y$ which induces H, H' and B on faces 3, 4 and 1 respectively, and which sends $\Delta[1] \times \Delta[1] \times \{t_0\}$ and $\{0\} \times \Delta[1] \times T$ to the base point y_0 of Y (t_0 is the base point of T), by IV, 5.2[1]. Passing to the quotient, α_1 defines a morphism $\alpha_2\colon \Delta[1] \wedge (\Delta[1] \times T) \to Y$ with which is canonically associated a morphism $\alpha_3\colon \Delta[1] \times T \to \mathcal{H}om_{\cdot}(\Delta[1], Y)$. It is then sufficient to choose α to be the equivalence class of the morphism $\Delta[1] \times T \to \Gamma f$ whose components are A and α_3.

1.3. Let $f\colon X \to Y$ be a morphism of pointed Kan complexes. By V, 4, the infinite sequence

$$(*) \quad \ldots \Omega^2 Y \xrightarrow{[\Omega q f]} \Omega \Gamma f \xrightarrow{[\Omega p f]} \Omega X \xrightarrow{[\Omega f]} \Omega Y \xrightarrow{[q f]} \Gamma f \xrightarrow{[p f]} X \xrightarrow{[f]} Y$$

is an r-exact sequence of "the category of pointed Kan complexes modulo homotopy". Moreover, if T is an arbitrary pointed complex and

[1] Apply IV, 5.2 to the Kan complex $\mathcal{H}om_{\cdot}(T, Y)$.

$a(T)\colon T\to T_K$ is an anodyne extension such that T_K is a Kan complex, the sets $.\overline{\Delta^\circ\mathscr{E}}(T, Y)$, $.\overline{\Delta^\circ\mathscr{E}}(T, X)$, $.\overline{\Delta^\circ\mathscr{E}}(T, \Gamma f)$, ... are canonically identified with the sets $.\overline{\Delta^\circ\mathscr{E}}(T_K, Y)$, $.\overline{\Delta^\circ\mathscr{E}}(T_K, X)$, $.\overline{\Delta^\circ\mathscr{E}}(T_K, \Gamma f)$, ... or with the sets $.\mathscr{H}(T, Y)$, $.\mathscr{H}(T, X)$, $.\mathscr{H}(T, \Gamma f)$ .... Hence we can also consider (∗) as an r-exact sequence of $.\overline{\Delta^\circ\mathscr{E}}$ or of $.\mathscr{H}$.

By 1.2, ΩY is the fibred product of the diagram of Fig. 76

$$\begin{array}{ccc} & & \mathscr{H}om.(\Delta[1], Y) \\ & & \downarrow{\scriptstyle e} \\ \Delta[0] = \mathscr{H}om.(\Delta[0], Y) & \to & \mathscr{H}om.(\dot\Delta[1], Y) \cong Y \end{array}$$

Fig. 76

This fibred product is obviously identified with $\mathscr{H}om.(\Omega, Y)$, where Ω is the simplicial circle (II, 2.5.2). Hence it follows from V, 3 that, for each pointed Kan complex, there is a canonical isomorphism from $.\overline{\Delta^\circ\mathscr{E}}(T, \mathscr{H}om.(\Omega, Y))$ onto $\Pi_1(\mathscr{H}om.(T, Y))$. It is easily seen that this isomorphism h can also be obtained from the canonical isomorphisms

$$.\Delta^\circ\mathscr{E}(T, \mathscr{H}om.(\Omega, Y)) \simeq .\Delta^\circ\mathscr{E}(T\wedge\Omega, Y) \simeq .\Delta^\circ\mathscr{E}(\Omega, \mathscr{H}om.(T, Y))$$

described in IV, 4.1.2 and from the isomorphism

$$.\overline{\Delta^\circ\mathscr{E}}(\Omega, \mathscr{H}om.(T, Y)) \simeq \Pi_1(\mathscr{H}om.(T, Y))$$

described in IV, 5.4.

1.4. Let $p\colon E\to B$ be a fibration, e_0 and b_0 base points of E and B such that $p(e_0)=b_0$, and F *the fibre of* p *over* b_0, i.e. the fibred product of the diagram

$$\begin{array}{ccc} & & E \\ & & \downarrow{\scriptstyle p} \\ \Delta[0] & \xrightarrow{\tilde b_0} & B \end{array}$$

This will be summed up by saying that the sequence

$$F \xrightarrow{i} E \xrightarrow{p} B$$

is a fibration (i denotes the inclusion of F into E).

Lemma: If B is a Kan complex, the morphism $j\colon F\to\Gamma p$ whose components are the inclusion i of F into E and the zero morphism from F to $\mathscr{H}om.(\Delta[1], B)$ is a homotopy equivalence.

F can be identified with a subcomplex of Γp by means of j; we will then construct a homotopy h, connecting a retraction of j with $\mathrm{Id}\,\Gamma p$, such that $eh_2 = ph_1$, if h_1 and h_2 are the components of h:

$$\begin{array}{ccc} \Delta[1]\times\Gamma p & \xrightarrow{h_2} & \mathscr{H}om.(\Delta[1], B) \\ {\scriptstyle h_1}\downarrow & & \downarrow{\scriptstyle e} \\ E & \xrightarrow{p} & B \end{array}$$

Fig. 77

In order to construct h_2, let us write $k\colon [1]\times[1]\to[1]$ for the increasing map which sends $(1, 0)$, $(0, 0)$, $(0, 1)$ to 0 and $(1, 1)$ to 1. The image $C(k)\colon \Delta[1]\times\Delta[1]\to\Delta[1]$ of k under the functor C of II, 5.1 defines then, by passing to the quotient, a morphism

$$l\colon \Delta[1]\wedge\Delta[1]\to\Delta[1]$$

which is a homotopy connecting the zero morphism with the identity of $\Delta[1]$ ($\Delta[1]$ is contractible!). Since $\Delta[1]$ is contractible, the same holds for $\mathcal{H}om.(\Delta[1], B)$: indeed, with the morphism

$$\mathcal{H}om.(l, B)\colon \mathcal{H}om.(\Delta[1], B)\to\mathcal{H}om.(\Delta[1]\wedge\Delta[1], B)$$

is canonically associated a morphism

$$l'\colon \Delta[1]\wedge\mathcal{H}om.(\Delta[1], B)\to\mathcal{H}om.(\Delta[1], B)$$

which connects the zero morphism and the identity of $\mathcal{H}om.(\Delta[1], B)$ (IV, 4.1.2). For h_2, we can then take the composition

$$\Delta[1]\times\Gamma p\xrightarrow{\text{can.}}\Delta[1]\wedge\Gamma p\xrightarrow{\Delta[1]\wedge\mathrm{pr}_2}\Delta[1]\wedge\mathcal{H}om.(\Delta[1], B)\xrightarrow{l'}\mathcal{H}om.(\Delta[1], B).$$

In order to construct h_1, we note that, by the Kan extension condition, there is a morphism h_1 such that $p\circ h_1=e\circ h_2$ and $h_1|\{1\}\times\Gamma p=\mathrm{Id}$. It follows that $p\circ(h_1|\{0\}\times\Gamma p)$ is zero, and hence that h_1 sends $\{0\}\times\Gamma p$ into the fibre. Since we can choose h_1 on $\Delta[1]\times F$ as we like, it is sufficient to require in addition that the restriction of h_1 to $\Delta[1]\times F$ should be the canonical projection onto F.

The above lemma implies the existence of an infinite sequence

$$\ldots\Omega^2 B\xrightarrow{\Omega([j]^{-1}[qf])}\Omega F\xrightarrow{[\Omega i]}\Omega E\xrightarrow{[\Omega p]}\Omega B\xrightarrow{[j]^{-1}[qf]}F\xrightarrow{[i]}E\xrightarrow{[p]}B$$

of the pointed homotopic category; this sequence is obviously isomorphic to the sequence (*) of 1.3; hence it is r-exact and depends functorially on p. In particular, each commutative diagram (Fig. 78)

$$\begin{array}{ccccc} F & \xrightarrow{i} & E & \xrightarrow{p} & B \\ \downarrow{\scriptstyle f} & & \downarrow{\scriptstyle e} & & \downarrow{\scriptstyle b} \\ F' & \xrightarrow{i'} & E' & \xrightarrow{p'} & B' \end{array}$$

Fig. 78

of $.\Delta^\circ\mathcal{E}$ whose rows are fibrations induces the following commutative diagram of $.\mathcal{H}$ (Fig. 79)

$$\begin{array}{ccccccc} \ldots\Omega B & \xrightarrow{[j]^{-1}[qf]} & F & \xrightarrow{[i]} & E & \xrightarrow{[p]} & B \\ \downarrow{\scriptstyle [\Omega b]} & & \downarrow{\scriptstyle [f]} & & \downarrow{\scriptstyle [e]} & & \downarrow{\scriptstyle [b]} \\ \ldots\Omega B' & \xrightarrow{[j']^{-1}[qf']} & F' & \longrightarrow & E' & \longrightarrow & B' \end{array}$$

Fig. 79

2. Cones

2.1. We intend now to show that the 2-category $.\mathfrak{Kan}$ also satisfies condition C°, dual to C (V, 6.2); in order to do this, let us consider first a morphism $f\colon X \to Y$ between arbitrary pointed sets. We will call *cone of f* (notation Cf) the amalgamated sum of the diagram

$$\begin{array}{ccc} X & \longrightarrow & Y \\ {\scriptstyle \varepsilon}\downarrow & & \\ \Delta[1] \wedge X & & \end{array}$$

where $\varepsilon\colon X \simeq \dot{\Delta}[1] \wedge X \to \Delta[1] \wedge X$ is the morphism induced by the inclusion of $\dot{\Delta}[1]$ into $\Delta[1]$; we also write if for the canonical injection of Y into the amalgamated sum Cf. Hence the canonical morphism from $\Delta[1] \wedge X$ to Cf is a homotopy relatively to base points connecting the zero morphism with $(if)f$; the equivalence class of this homotopy is a 2-morphism of $((.\Delta^\circ\mathscr{E}))$, which will be written kf.

Since $((.\Delta^\circ\mathscr{E}))$ is a 2-category satisfying conditions A and B of V, 1, the triple (Cf, if, kf) defines, for each pointed simplicial set T, a functor from the groupoid $(Cf)_T$ to the groupoid $\Gamma(f_T)$ (V, 6.2). Our triple does not satisfy completely condition C°; however, we have the following result:

Lemma: If T is a pointed Kan complex, the functor from $(Cf)_T$ to $\Gamma(f_T)$ defined by the triple (Cf, if, kf), is connected.

Let $i\colon Y \to T$ be a 1-morphism and $k\colon 0 \to i \cdot f$ a 2-morphism. Since T is a Kan complex, there is a morphism $K\colon \Delta[1] \wedge X \to T$ whose class is k. If we write r for the morphism from the amalgamated sum Cf to T whose components are K and i, we then have $r(if) = i$ and $r * (kf) = k$.

Now let $r, r'\colon Cf \rightrightarrows T$ be two 1-morphisms, and let $i = r(if)$, $i' = r'(if)$, $k = r * (kf)$ and $k' = r' * (kf)$. We must show that, for each 2-morphism $a\colon i \to i'$ such that $(a * f) \circ k = k'$, there is a 2-morphism $\alpha\colon r \to r'$ such that $a = \alpha * (if)$. The construction of α is similar to that given in 1.2.2. This completes the proof.

2.2. *We are now ready to verify condition C° in the 2-category* $.\mathfrak{Kan}$: suppose then that X and Y are Kan complexes; unfortunately, in that case, *Cf is not general a Kan complex.* However, take an anodyne extension $a\colon Cf \to C'f$ such that $C'f$ is a Kan complex, and let $i'f = a(if)$ and $k'f = a * (kf)$.

If T is a pointed Kan complex, the functor $(C'f)_T \to \Gamma(f_T)$ defined by the triple $(C'f, i'f, k'f)$ is obviously the composition of the functor $a_T\colon (C'f)_T \to (Cf)_T$ induced by a, with the functor defined by the triple (Cf, if, kf). Since the latter is connected by 2.1, it is sufficient to verify that a_T is connected: since each morphism from Cf to T can be extended to $C'f$, a_T induces a surjection on objects. On the other hand, let j, j': $C'f \rightrightarrows T$ be two objects of $(C'f)_T$ and let β: $ja \to j'a$ be a morphism of the groupoid $(Cf)_T$.

Since the extension

$$(\Delta[1] \times Cf) \cup (\dot{\Delta}[1] \times C'f) \to \Delta[1] \times C'f$$

is anodyne by IV, 2.2, each morphism B: $\Delta[1] \times Cf \to T$ whose equivalence class in $(Cf)_T$ is β (see IV, 5.2), can be extended to a homotopy A: $\Delta[1] \times C'f \to T$ between j and j'. The image α of A in $(C'f)_T$ is such that $\beta = \alpha * a$; this complete the proof.

The preceding remarks and paragraph V, 6 imply in particular that *$\Omega^2 Y$ is a $.\mathscr{H}$-abelian group for each pointed Kan complex Y.*

2.3. Let us return now to *arbitrary pointed simplicial sets X and Y.* We will call *suspension* of X (notation: ΣX) the pointed complex $\Omega \wedge X$, where Ω is the simplicial circle (II, 2.5.2). Thus the suspension of X is simply the cone of the zero morphism $X \to 0$.

If f: $X \to Y$ is a morphism of pointed complexes, the commutative diagram

$$\begin{array}{ccc} X & \xrightarrow{f} & Y \\ {\scriptstyle \mathrm{Id}\,X}\downarrow & & \downarrow{\scriptstyle 0^Y} \\ X & \xrightarrow{0^X} & 0 \end{array}$$

induces a morphism jf from Cf to the cone ΣX of 0^X. Hence, if we consider Cf as an amalgamated sum, the components of jf are the canonical morphism from $\Delta[1] \wedge X$ to ΣX and the zero morphism from Y to ΣX.

Theorem: For each morphism f: $X \to Y$ of $.\Delta^\circ\mathscr{E}$, the infinite sequence

$$\text{(i)} \qquad X \xrightarrow{f} Y \xrightarrow{if} Cf \xrightarrow{jf} \Sigma X \xrightarrow{\Sigma f} \Sigma Y \xrightarrow{\Sigma if} \Sigma Cf \xrightarrow{\Sigma jf} \Sigma^2 X \ldots$$

induces a l-exact sequence of the pointed homotopic category.

Recall (V, 6.3.2) that our theorem means that, for each pointed complex T, the infinite sequence of pointed sets

$$\text{(i)}_T \quad \ldots .\mathscr{H}(\Sigma Y, T) \xrightarrow{.\mathscr{H}(\Sigma f, T)} .\mathscr{H}(\Sigma X, T) \xrightarrow{.\mathscr{H}(jf, T)} .\mathscr{H}(Cf, T) \xrightarrow{.\mathscr{H}(if, T)} .\mathscr{H}(Y, T) \xrightarrow{.\mathscr{H}(f, T)} .\mathscr{H}(X, T$$

is exact. Note also that the sequences (i) and (i)_T depend "functorially" on f.

We will prove this theorem by comparing the sequence (i) to sequences which we know to be l-exact by V, 6. In order to do this, consider the Kan envelope $(e, ?_K)$ of IV, 4.4, and let us write $C_K f$ instead of $(Cf)_K$,

$\Sigma_K Z$ instead of $(\Sigma Z)_K$, $i_K f$ instead of $e(Cf)\cdot(if)$ and $j_K f$ instead of $(jf)_K$. We then have the commutative diagram of Fig. 80

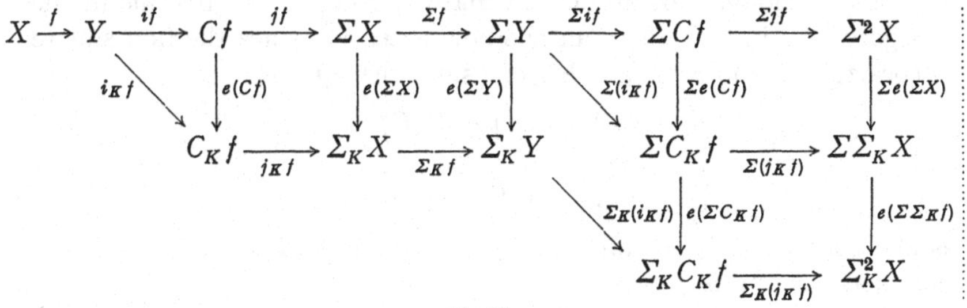

Fig. 80

Since all vertical arrows are anodyne extensions by IV, 4.3, the sequence (i) is isomorphic, in the pointed homotopic category, to the infinite sequence

(ii) $X \xrightarrow{f} Y \xrightarrow{i_K f} C_K f \xrightarrow{j_K f} \Sigma_K X \xrightarrow{\Sigma_K f} \Sigma_K Y \xrightarrow{\Sigma_K(i_K f)} \Sigma_K C_K f \ldots .$

When X and Y are Kan complexes, the latter sequence is l-exact by 2.2 and V, 6.3.2.

If we suppose only that X is a Kan complex, we have a commutative diagram (Fig. 81)

(iii)

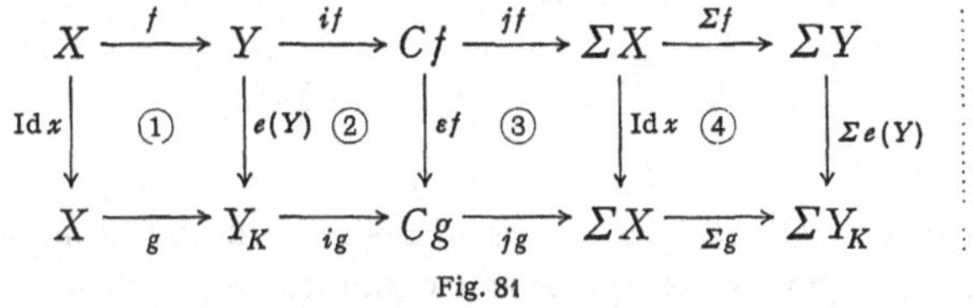

Fig. 81

where g is the composition $e(Y)f$ and where εf is induced by $e(Y)$. Since the square 2 is obviously cocartesian, εf and all vertical arrows of the diagram are anodyne extensions (IV, 4.3). The rows are then isomorphic in the pointed homotopic category, which takes us back to the preceding case.

Finally, when we make no hypothesis on X and Y, we have a commutative diagram (Fig. 82)

(iv)

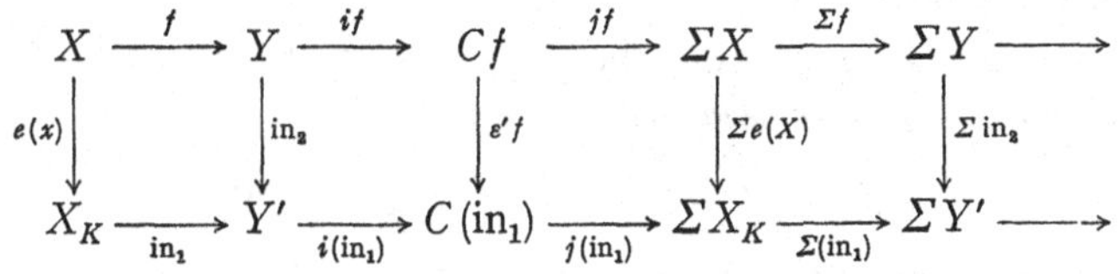

Fig. 82

where Y' is the amalgamated sum of X_K and Y under X, in_1 and in_2 are the canonical morphisms, and where $\varepsilon' f$ is induced by in_2; all vertical arrows are anodyne extensions, so that the rows are isomorphic in the category $.\mathscr{H}$: in fact, in_2 is an anodyne extension since the first square is cocartesian, and $\varepsilon' f$ is another one since the square

$$\begin{array}{ccc} X \wedge \Delta[1] & \longrightarrow & Cf \\ {\scriptstyle e(K)\wedge I}\downarrow & & \downarrow{\scriptstyle \varepsilon' f} \\ X_K \wedge \Delta[1] & \longrightarrow & C(\mathrm{in}_1) \end{array}$$

is also cocartesian. This takes us back to the preceding case, and hence completes the proof.

2.4. We keep here the same notations as in 2.3, and we suppose moreover that f is *a monomorphism of* $.\Delta^\circ\mathscr{E}$. We write then Y/X for the amalgamated sum of the following diagram of $.\Delta^\circ\mathscr{E}$:

$$\begin{array}{ccc} X & \xrightarrow{f} & Y \\ {\scriptstyle 0^X}\downarrow & & \\ o & & \end{array}$$

Since 0^X is the composition of the morphism $\varepsilon\colon X \to \Delta[1] \wedge X$ of 2.1 with the zero morphism $\Delta[1] \wedge X \to o$, Y/X is also identified with $Cf/(\Delta[1] \wedge X)$; in other words, the diagram of Fig. 83

$$\begin{array}{ccccc} X & \xrightarrow{\varepsilon} & \Delta[1] \wedge X & \longrightarrow & o \\ {\scriptstyle f}\downarrow & & \downarrow & & \downarrow \\ Y & \xrightarrow{if} & Cf & \xrightarrow{\varrho} & Y/X \end{array}$$

Fig. 83

is commutative. (Here, ϱ is the morphism from the amalgamated sum to Y/X whose components are the canonical projection p from Y to Y/X and the zero morphism.) If l denotes the morphism $\Delta[1] \wedge \Delta[1] \to \Delta[1]$ of 1.4, we write $h\colon \Delta[1] \wedge \Delta[1] \wedge X \to \Delta[1] \wedge X$ for the morphism $l \wedge \mathrm{Id}\, X$. This morphism is a homotopy relatively to base points connecting the zero morphism with the identity of $\Delta[1] \wedge X$. In other words, $\Delta[1] \wedge X$ is contractible. Applying lemma 2.4.1 below to the monomorphism $\Delta[1] \wedge X \to Cf$, we get

Proposition: If $f\colon X \to Y$ is a monomorphism of pointed simplicial sets, the canonical morphism from Cf to Y/X has an invertible image in the pointed homotopic category.

If f is a monomorphism, the sequence (i) of 2.3 is thus isomorphic (in $.\mathscr{H}$) to the *infinite sequence*

$$X \xrightarrow{f} Y \xrightarrow{p} Y/X \xrightarrow{if\circ\varrho^{-1}} \Sigma X \xrightarrow{\Sigma f} \Sigma Y \ldots.$$

This sequence is a l-exact sequence of $.\mathscr{H}$.

2.4.1. *Lemma: Let $u: U \to V$ be a monomorphism of simplicial sets. If U is contractible, the canonical projection p of V onto V/U has an invertible image in $.\mathcal{H}$.*

The proof is similar to that of lemma V, 7.5. Since U is contractible, there is a composed homotopy $h: I_n \times U \to U$ (relatively to base points) connecting the zero morphism with $\mathrm{Id}\, U$. Let then g be the morphism from $(I_n \times U) \cup (\{0\} \times V)$ to V whose restriction to $I_n \times U$ is h, and which is induced by $\mathrm{Id}\, V$ on $\{0\} \times V$ (we identify U to a subcomplex of V by means of u). Consider then the commutative diagram of $\Delta^\circ \mathcal{E}$ of Fig. 84

$$\text{(i)} \qquad \begin{array}{ccc} (I_n \times U) \cup (\{0\} \times V) & \xrightarrow{g} & V \\ {\scriptstyle \text{incl.}}\downarrow & & \downarrow{\scriptstyle \sigma} \\ I_n \times V & \xrightarrow{G} & V' \end{array}$$

Fig. 84

where V' is the amalgamated sum of the full line diagram, and where σ and G are the canonical morphisms. In this diagram, σ is an anodyne extension, and G is a homotopy (relatively to base points) connecting σ with a morphism f from V to V'; finally f factors through p by means of a morphism $r: V/U \to V'$. Let then q be the morphism of $.\mathcal{H}$ defined by the following diagram of $.\overline{\Delta^\circ \mathcal{E}}$ (see I, 2.3)

$$V/U \xrightarrow{r} V' \xleftarrow{\sigma} V.$$

Since we have $f = r \circ p$, the composition $q \circ p$ in $.\mathcal{H}$ is defined by the diagram

$$V \xrightarrow{f} V' \xleftarrow{\sigma} V.$$

Since, by construction, $[f] = [\sigma]$ in the category $.\overline{\Delta^\circ \mathcal{E}}$, we have $q \circ p = \mathrm{Id}\, V$ in the pointed homotopic category.

Let us now write p' for the projection of V' onto V'/U and $\sigma': V/U \to V'/U$ for the morphism induced by σ. The square

$$\text{(ii)} \qquad \begin{array}{ccc} V & \xrightarrow{p} & V/U \\ {\scriptstyle \sigma}\downarrow & & \downarrow{\scriptstyle \sigma'} \\ V' & \xrightarrow[p']{} & V'/U \end{array}$$

is cocartesian, and hence σ' is an anodyne extension; but in the category $.\mathcal{H}$, the morphism $p \circ q$ is represented by the diagram

$$V/U \xrightarrow{p' \circ r} V'/U \xleftarrow{\sigma'} V/U$$

(see I, 2.3): in order to show that $p \circ q = \mathrm{Id}(V/U)$ in the pointed homotopic category, it is sufficient to find a homotopy H (relatively to base points) connecting σ' with $p' \circ r$: we note then that G induces, by passing to the quotients, a morphism from $I_n \times (V/U)$ to V'/U, and this is the homotopy we were looking for.

3. Homotopy Groups

3.1. Let X be an arbitrary pointed complex. By 2.3, the suspension ΣX of X is simply the contracted product $\Omega \wedge X$. Moreover, the functor $Y \rightsquigarrow Y \wedge X$ from $.\mathscr{H}$ to $.\mathscr{H}$ is left adjoint to the functor $Z \rightsquigarrow \mathscr{H}om.(X, Z_K)$ described in IV, 4.3.1. It follows that the functor $Y \rightsquigarrow Y \wedge X$ commutes with direct sums, and hence that the canonical morphism

$$(\Omega \wedge X) \vee (\Omega \wedge X) \to (\Omega \vee \Omega) \wedge X$$

is an isomorphism. If we compose this isomorphism with the morphism $\varphi \wedge X$ deduced from IV, 4.5, we give to $\Omega \wedge X$ a cogroup structure of the category $.\mathscr{H}$.

When $x\colon X \to X'$ is a morphism of $.\mathscr{H}$, the induced morphism $\Sigma x\colon \Sigma X \to \Sigma X'$ is obviously a cogroup homomorphism. In the sequence (i) of Theorem 2.3, in particular, the morphisms Σf, $\Sigma i f$, $\Sigma j f$ ... are cogroup homomorphisms.

Finally, suppose that X is a cogroup of $.\mathscr{H}$ and let $\delta\colon X \to X \vee X$ be the comultiplication of this cogroup. Then the cogroup structures of $\Omega \wedge X$ which are deduced from $\varphi \wedge X$ and $\Omega \wedge \delta$ coincide, and make $\Omega \wedge X$ an abelian cogroup. In the sequence (i) of theorem 2.3, in particular, the objects $\Sigma^2 X$, $\Sigma^2 Y$, ... are abelian cogroups.

3.2. Let T be a pointed complex, and set

$$\Pi_0 T = .\mathscr{H}(\dot{\Delta}[1], T),\ \Pi_1 T = .\mathscr{H}(\Sigma\dot{\Delta}[1], T), \ldots, \Pi_n T = .\mathscr{H}(\Sigma^n\dot{\Delta}[1], T).$$

By 3.1, the sets $\Pi_n T$ have a group structure for $n \geqq 1$: we will say that $\Pi_n T$ is *the n-th homotopy group of* T; this group is abelian for $n \geqq 2$.

The set $\Pi_0 T$ is simply the set of connected components of T. Let us consider then $\Pi_1 T$: for each pointed complex Y, we can identify Y with $Y \wedge \dot{\Delta}[1]$ by means of the following composed morphism of $.\Delta^\circ\mathscr{E}$:

$$Y \simeq Y \times \{1\} \xrightarrow{\text{incl.}} Y \times \dot{\Delta}[1] \xrightarrow{\text{can.}} Y \wedge \dot{\Delta}[1].$$

In particular, $\Sigma\dot{\Delta}[1]$ is simply the simplicial circle Ω. By IV, 5, $\Pi_1 T$ is thus identified with the Poincaré group of T at its base point.

More generally, $\Sigma^n\dot{\Delta}[1]$ is identified with the contracted product $\wedge^n\Omega$ of n copies of Ω. If p is the canonical projections of $\Delta[1]$ onto Ω, the composition

$$q_n\colon \Delta[1] \cdots \times \Delta[1] \xrightarrow{p \times \cdots \times p} \Omega \times \cdots \times \Omega \xrightarrow{\text{can.}} \Omega \wedge \cdots \wedge \Omega$$

is an epimorphism of $.\Delta^\circ\mathscr{E}$. Besides, if follows from the definition of Ω as cokernel of the pair of morphisms

$$\Delta(\partial_1^0), \Delta(\partial_1^1)\colon \Delta[0] \rightrightarrows \Delta[1],$$

that the square

$$\begin{array}{ccc} A & \xrightarrow{\text{incl.}} & \Delta[1]^n \\ \downarrow & & \downarrow{\scriptstyle q_n} \\ \Delta[0] & \longrightarrow & \wedge^n \Omega \end{array}$$

where $A = \bigcup_{n=p+q+1} \Delta[1]^p \times \dot{\Delta}[1] \times \Delta[1]^q$, is cocartesian. *Hence the pointed complex $\wedge^n \Omega$ is obtained from the n-dimensional cube by contracting the "boundary" of the cube to a point.*

3.3 If X is a pointed complex and T a *pointed Kan complex*, we have, by IV, 4.2 and IV, 4.3.1, a functor isomorphism

$$.\mathscr{H}(\Sigma X, T) \simeq .\mathscr{H}(X, \Omega T)$$

where $\Omega T = \mathscr{H}om.(\Omega, T)$ (see 1.3). Hence, for each $n \geqq 1$, we have functor isomorphisms

$$.\mathscr{H}(\Sigma^n X, T) \simeq .\mathscr{H}(\Sigma^{n-1} X, \Omega T) \simeq \cdots \simeq .\mathscr{H}(X, \Omega^n T).$$

For each morphism $f\colon Z \to T$ of pointed Kan complexes, the exact sequence

$$\ldots .\mathscr{H}(X, \Omega T) \xrightarrow{.\mathscr{H}(X, qf)} .\mathscr{H}(X, \Gamma f) \xrightarrow{.\mathscr{H}(X, pf)} .\mathscr{H}(X, Z) \xrightarrow{.\mathscr{H}(X, f)} .\mathscr{H}(X, T)$$

induces then an isomorphic exact sequence

$$\ldots .\mathscr{H}(\Sigma X, Z) \xrightarrow{.\mathscr{H}(\Sigma X, f)} .\mathscr{H}(\Sigma X, T) \longrightarrow .\mathscr{H}(X, \Gamma f) \xrightarrow{.\mathscr{H}(X, pf)} .\mathscr{H}(X, Z) \xrightarrow{.\mathscr{H}(X, f)} .\mathscr{H}(X, T).$$

In particular, if X is equal to $\dot{\Delta}[1]$, we have the exact sequence of homotopy groups:

$$\ldots \Pi_1 Z \xrightarrow{\Pi_1 f} \Pi_1 T \longrightarrow \Pi_0 \Gamma f \xrightarrow{\Pi_0 pf} \Pi_0 Z \xrightarrow{\Pi_0 f} \Pi_0 T.$$

Similarly, consider a pointed Kan complex B and a fibration (1.5)

$$F \xrightarrow{i} E \xrightarrow{p} B$$

The exact sequence

$$\ldots \Omega B \longrightarrow F \xrightarrow{[i]} E \xrightarrow{[p]} B$$

of 1.5 and the functor isomorphisms $\Pi_n T \simeq .\mathscr{H}(\Delta[1], T)$ define an exact sequence of the homotopy groups of a fibration

$$\ldots \Pi_2 B \longrightarrow \Pi_1 F \xrightarrow{\Pi_1 i} \Pi_1 E \xrightarrow{\Pi_1 p} \Pi_1 B \longrightarrow \Pi_0 F \xrightarrow{\Pi_0 i} \Pi_0 E \xrightarrow{\Pi_0 p} \Pi_0 B.$$

This exact sequence depends "functorially" on p.

3.4. We will now give another definition of the homotopy groups, which will be needed at the end of this chapter. We will see indeed that the pointed complex $\wedge^n \Omega$ is isomorphic in $.\mathscr{H}$ to the complex $\Delta[n]/\dot{\Delta}[n]$

obtained from $\Delta[n]$ by contracting its boundary to a point. Thus, for each pointed complex T, we will have an isomorphism

$$\Pi_n T \simeq .\mathcal{H}(\Delta[n]/\dot{\Delta}[n], T)$$

which we will use later.

Indeed, it is clear that, in the category $.\Delta^\circ\mathcal{E}$, the morphism $\Delta[\partial^0_{n+1}]$: $\Delta[n]\to\Delta[n+1]$ induces an isomorphism

$$\Delta[n]/\dot{\Delta}[n]\simeq\dot{\Delta}[n+1]/\Lambda^0[n+1].$$

Since $\Lambda^0[n+1]$ is contractible (a homotopy connecting the identity of $\Lambda^0[n+1]$ with the zero morphism is given by $C(v^0_{n+1})$, IV, 2.1.3), we then have an isomorphism

$$\Delta[n]/\dot{\Delta}[n]\simeq\dot{\Delta}[n+1]$$

in the pointed homotopic category (2.4.1).

Let us now write f for the canonical injection of $\dot{\Delta}[n]$ into $\Delta[n]$. It follows from the definitions that the two following squares of $.\Delta^\circ\mathcal{E}$ are cocartesian:

$$\begin{array}{ccc} \dot{\Delta}[n] & \longrightarrow & \Delta[1]\wedge\dot{\Delta}[n] \\ {\scriptstyle f}\downarrow & & \downarrow \\ \Delta[n] & \xrightarrow[i_f]{} & Cf \end{array} \qquad\qquad \begin{array}{ccc} \Delta[n] & \xrightarrow{i_f} & Cf \\ \downarrow & & \downarrow \\ \Delta[0] & \longrightarrow & Cf/\Delta[n] \end{array}$$

Fig. 85

Hence the same holds for the composition of these two squares (Fig. 86)

$$\begin{array}{ccc} \dot{\Delta}[n] & \longrightarrow & \Delta[1]\wedge\dot{\Delta}[n] \\ \downarrow & & \downarrow \\ \Delta[0] & \longrightarrow & Cf/\Delta[n] \end{array}$$

Fig. 86

In other words, $Cf/\Delta[n]$ is the cone of the zero morphism $\dot{\Delta}[n]\to 0$, i.e. we have $Cf/\Delta[n]\simeq\Sigma\dot{\Delta}[n]$ in the category $.\Delta^\circ\mathcal{E}$.

But, in the pointed homotopic category, we have the following isomorphism:

– $Cf/\Delta[n]\simeq Cf$, for $\Delta[n]$ is contractible
– $Cf\simeq\Delta[n]/\dot{\Delta}[n]$ by 2.3
– $\dot{\Delta}[n]\simeq\Delta[n-1]/\dot{\Delta}[n-1]$ $(n>1)$.

We deduce an isomorphism $(n>1)$:

$$\Delta[n]/\dot{\Delta}[n]\simeq\Sigma(\Delta[n-1]/\dot{\Delta}[n-1]).$$

Since by definition, $\Omega = \Delta[1]/\dot{\Delta}[1] = \Sigma\dot{\Delta}[1]$, it follows from the above isomorphism that

$$\Delta[n]/\dot{\Delta}[n] \simeq \Sigma^{n-1}\Omega = \Sigma^n\dot{\Delta}[1]. \qquad \text{Q.E.D.}$$

4. Generalities on Fibrations

4.1. Let $p\colon E \to B$ and $p'\colon E' \to B'$ be two morphisms of a category $\mathscr{C}$. A morphism from p to p' (see V, 4) is a pair (u, v) formed by morphisms $u\colon E \to E'$ and $v\colon B \to B'$ such that $p'u = vp$. These morphisms are composed in an obvious way, so that we will be able to speak of the category of morphisms of $\mathscr{C}$, written $\mathscr{A}r\,\mathscr{C}$. In particular, if p and p' are two isomorphic objects of this category, we will say that p and p' are *isomorphic morphisms* of $\mathscr{C}$.

4.1.1. For example, consider, the case of the category $\Delta^\circ\mathscr{E}$. By definition, $(\mathscr{A}r\,\Delta^\circ\mathscr{E})(p, p')$ is the fibred product of the diagram of sets of Fig. 87

$$\begin{array}{ccc} & & \Delta^\circ\mathscr{E}(E, E') \\ & & \Big\downarrow{\scriptstyle \Delta^\circ\mathscr{E}(E, p')} \\ \Delta^\circ\mathscr{E}(B, B') & \xrightarrow{\Delta^\circ\mathscr{E}(p, B')} & \Delta^\circ\mathscr{E}(E, B') \end{array}$$

Fig. 87

This leads to write $\mathscr{H}om(p, p')$ for the fibred product of the following diagram of $\Delta^\circ\mathscr{E}$ (Fig. 88)

$$\begin{array}{ccc} & & \mathscr{H}om(E, E') \\ & & \Big\downarrow{\scriptstyle \mathscr{H}om(E, p')} \\ \mathscr{H}om(B, B') & \xrightarrow{\mathscr{H}om(p, B')} & \mathscr{H}om(E, B') \end{array}$$

Fig. 88

Thus an n-simplex of $\mathscr{H}om(p, p')$ is a pair (h, k) which makes the following diagram commutative:

$$\begin{array}{ccc} \Delta[n] \times E & \xrightarrow{h} & E' \\ \Big\downarrow{\scriptstyle \Delta[n]\times p} & & \Big\downarrow{\scriptstyle p'} \\ \Delta[n] \times B & \xrightarrow{l} & B' \end{array}$$

The face and degeneracy operators are obvious; moreover, $\mathscr{H}om(p, p')$ is a Kan complex if p' is a fibration (IV, 3.1.1 and IV, 3.1.2) and B' is a Kan complex.

4.1.2. Two morphisms from p to p' are said to be *homotopic* if they belong to the same connected component of $\mathscr{H}om(p, p')$. The set of these connected components will be written $\Pi_0(p, p')$. If $p''\colon E'' \to B''$ is a third morphism of $\Delta^\circ\mathscr{E}$, it is clear that the composition map

$$(\mathscr{A}r\,\Delta^\circ\mathscr{E})(p, p') \times (\mathscr{A}r\,\Delta^\circ\mathscr{E})(p', p'') \to (\mathscr{A}r\,\Delta^\circ\mathscr{E})(p, p'')$$

is compatible with the homotopy relations, and induces a map

$$\Pi_0(p, p') \times \Pi_0(p', p'') \to \Pi_0(p, p'').$$

Thus we define the category of morphisms of $\Delta^\circ\mathscr{E}$ modulo homotopy: its objects are the morphisms of $\Delta^\circ\mathscr{E}$, its sets of morphisms are the sets $\Pi_0(p, p')$. If p and p' are two isomorphic objects of this category, we will say that *p and p' have the same homotopy type.*

4.1.3. Suppose now that E is a subcomplex of E', that B is a subcomplex of B', and that $p\colon E \to B$ is induced by p'. Then we say that p is a *deformation retract* of p' if there is a pair (h, k) such that the square

$$\begin{array}{ccc} \Delta[1] \times E' & \xrightarrow{h} & E' \\ \downarrow{\scriptstyle \Delta[1]\times p'} & & \downarrow{\scriptstyle p'} \\ \Delta[1] \times B' & \xrightarrow{k} & B' \end{array}$$

is commutative, that the restrictions of h and k to $\{0\} \times E' \simeq E'$ and $\{0\} \times B' \simeq B'$ are the identity morphisms of E' and B', and that the restrictions of h and k to $\Delta[1] \times E$ and $\Delta[1] \times B$ factor through the canonical projections onto E and B. Such a pair (h, k) is a 1-simplex of $\mathscr{H}om(p', p')$; we will say that it is a *retracting deformation* of p' onto p.

If p is a deformation retract of p', it is clear that p and p' have the same homotopy type (4.1.2).

4.2. Let us return now to the notations of 4.1.1, supposing that *the bases B and B' coincide.* We write then $\mathscr{H}om_B(p, p')$ or $\mathscr{H}om_B(E, E')$ for the fibre of the canonical map

$$\mathscr{H}om(p, p') \to \mathscr{H}om(B, B)$$

over the identity morphism of B (1.4; note that $\mathrm{Id}\,B$ is a 0-simplex of $\mathscr{H}om(B, B)$). If p' is a fibration, $\mathscr{H}om_B(p, p')$ is a Kan complex by IV, 3.1.1 and V, 3.1.2.

The n-simplices of $\mathscr{H}om_B(p, p')$ correspond then to the morphisms $h\colon \Delta[n] \times E \to E'$ of $\Delta^\circ\mathscr{E}$ such that the triangle

$$\begin{array}{ccc} \Delta[n] \times E & \xrightarrow{h} & E' \\ & {\scriptstyle p\circ \mathrm{pr}_2}\searrow \quad \swarrow{\scriptstyle p'} & \\ & B & \end{array}$$

commutes (pr_2 is the canonical projection onto the second factor); the 0-simplices, in particular, are identified with the morphisms $u\colon E \to E'$ of $\Delta^\circ\mathscr{E}$ such that $p'u = p$. These morphisms are composed in an obvious way and allow us to define the category of complexes over B, whose objects are the morphisms p of $\Delta^\circ\mathscr{E}$ with range B. We say that p and p' (or sometimes E and E') are *isomorphic relatively* to B, if they are isomorphic objects of this category.

4.2.1. Let $p\colon E\to B$ and $p'\colon E'\to B$ be two morphisms of $\Delta^\circ\mathscr{E}$ with the same range B. Two morphisms $u, u'\colon E \rightrightarrows E'$, such that $p=p'u=p'u'$, are called *homotopic relatively to* B if they belong to the same connected component of $\mathscr{H}om_B(p, p')$. The set of these connected components will be written $\Pi_B^\circ(p, p')$. If $p''\colon E''\to B$ is a third morphism with range B, the composition map

$$\Delta^\circ\mathscr{E}(E, E')\times\Delta^\circ\mathscr{E}(E', E'')\to\Delta^\circ\mathscr{E}(E, E'')$$

induces a map

$$\Pi_B^\circ(p, p')\times\Pi_B^\circ(p', p'')\to\Pi_B^\circ(p, p'').$$

This allows us to define *the category of complexes over B modulo homotopy*: the objects are the morphisms of $\Delta^\circ\mathscr{E}$ with range B; the sets of morphisms are the sets $\Pi_B^\circ(p, p')$ .... If p and p' are two isomorphic objects of this category, we will say that p and p' have *the same homotopy type relatively to* B.

4.2.2. With the notations of 4.2.1, suppose now that E is a subcomplex of E' and that p is the restriction of p' to E. We say that p (or E) is a *deformation retract of* p' (or E') *relatively to* B if there is a morphism $h\colon \Delta[1]\times E'\to E'$ such that $p'\circ h=p'\circ \mathrm{pr}_2$ and that the restriction of h to $E'\simeq\{0\}\times E'$ (resp. to $\Delta[1]\times E$) is the identity of E' (resp. the canonical projection onto E); such an h will be called a retracting deformation of E' onto E relatively to B.

If p is a deformation retract of p' relatively to B, p and p' have the same homotopy type relatively to B.

4.3. Let us return now to the notations of 4.1.1 and to the definition of $\mathscr{H}om(p, p')$. The commutative square (Fig. 89)

$$\begin{array}{ccc}
\mathscr{H}om(B, E') & \xrightarrow{\mathscr{H}om(p,\,E')} & \mathscr{H}om(E, E') \\
\Big\downarrow{\scriptstyle \mathscr{H}om(B,\,p')} & & \Big\downarrow{\scriptstyle \mathscr{H}om(E,\,p')} \\
\mathscr{H}om(B, B') & \xrightarrow[\ ]{\mathscr{H}om(p,\,B')} & \mathscr{H}om(E, B')
\end{array}$$

Fig. 89

induces then a morphism $p/p'\colon \mathscr{H}om(B, E')\to\mathscr{H}om(p, p')$ whose components are $\mathscr{H}om(p, E')$ and $\mathscr{H}om(B, p')$.

Proposition: If $i\colon Y\to X$ is a monomorphism and $p\colon E\to B$ a fibration, the morphism $i/p\colon \mathscr{H}om(X, E)\to\mathscr{H}om(i, p)$ is a fibration.

Consider a commutative square (Fig. 90).

$$\begin{array}{ccc} U & \xrightarrow{a} & \mathcal{H}om(X, E) \\ {\scriptstyle u}\downarrow & & \downarrow{\scriptstyle i/p} \\ V & \xrightarrow{b} & \mathcal{H}om(i, p) \end{array}$$

Fig. 90

where u is an anodyne extension; we even suppose that u is an inclusion. Let then $b_1: V \to \mathcal{H}om(Y, E)$ and $b_2: V \to \mathcal{H}om(X, B)$ be the components of b; let $a': U \times X \to E$, $b'_1: V \times Y \to E$ and $b'_2: V \times X \to B$ be the morphisms canonically associated, by adjunction, with a, b_1 and b_2 (II, 2.5.3). The relation $\mathcal{H}om(i, E) \circ a = b_1 \circ u$ implies $a' \circ (U \times i) = b'_1 \circ (u \times Y)$, so that a' and b'_1 coincide on the intersection $U \times Y$ of their domains. Hence there is a square (Fig. 91)

$$\begin{array}{ccc} V \times Y \cup U \times X & \xrightarrow{c'} & E \\ \downarrow{\scriptstyle \text{incl.}} & & \downarrow{\scriptstyle p} \\ V \times X & \xrightarrow{b'_2} & B \end{array}$$

Fig. 91

such that the restrictions of c' to $V \times Y$ and $U \times X$ are equal respectively to b'_1 and a'. Moreover, the relations

$$\mathcal{H}om(Y, p) \circ b_1 = \mathcal{H}om(i, B) \circ b_2 \quad \text{and} \quad \mathcal{H}om(X, p) \circ a = b_2 \circ u$$

imply that $p b'_1 = b'_2\ (V \times i)$ and $p a' = b'_2 (u \times X)$. Hence our square is commutative. Since the vertical arrow on the left is an anodyne extension by IV, 2.2, there is a morphism $d': V \times X \to E$ whose restriction to $V \times Y \cup U \times X$ is c' and such that $b'_2 = p d'$. Then the morphism $d: V \to \mathcal{H}om(X, E)$, canonically associated with d (II, 2.5.3), satisfies the required equalities $du = a$ and $(i/p)\, d = b$.

When Y is the empty complex, i/p is identified with $\mathcal{H}om(X, p)$ and we get IV, 3.1.2 again; when B is equal to $\Delta[0]$, i/p is identified with $\mathcal{H}om(i, E)$ and we get V, 3.1.3.

4.3.1. We say that a morphism $p: E \to B$ of $\Delta^\circ \mathcal{E}$ has the *path lifting property* if, for each commutative square of the form

$$\begin{array}{ccc} \Delta[0] & \xrightarrow{u} & E \\ {\scriptstyle j}\downarrow & & \downarrow{\scriptstyle p} \\ \Delta[1] & \xrightarrow[v]{} & B \end{array}$$

there is a morphism $w: \Delta[1] \to E$ such that $wj = u$ and $pw = v$.

Corollary: Let $p: E \to B$ be a morphism of $\Delta^\circ \mathcal{E}$. Then the following statements are equivalent:

(i) *p is a fibration;*

(ii) *for each monomorphism $i\colon Y\to X$, the morphism*

$$i/p\colon \mathcal{H}om(X, E)\to \mathcal{H}om(i, p)$$

has the path lifting property.

The implication (i)$\Rightarrow$(ii) follows from proposition 4.3. On the other hand, statement (ii) means that, for each subcomplex Y of a complex X, and for each commutative square

$$\begin{array}{ccc} \Delta[1]\times Y\cup\{e\}\times X & \xrightarrow{a} & E \\ \big\downarrow{\scriptstyle \text{incl.}} & & \big\downarrow{\scriptstyle p} \\ \Delta[1]\times X & \xrightarrow{b} & B \end{array} \qquad e=0,1$$

there is a morphism $c\colon \Delta[1]\times X\to E$ whose restriction to $\Delta[1]\times Y\cup\{e\}\times X$ is a and which is such that $pc=b$. This implies (i) by IV, 3.1.

4.3.2. Let us return now to the notations of 4.3 in the particular case where $X=\Delta[n]$ and $Y=\dot{\Delta}[n]$ (II, 3.6; then Y is empty if $n=0$). A point (a, b) of $\mathcal{H}om(i, p)$ is then given by a singular simplex $b\colon \Delta[n]\to B$ of the base, and a "lifting" $a\colon \dot{\Delta}[n]\to E$ of the restriction of b to $\dot{\Delta}[n]$ (hence we have $p\circ a=b|\dot{\Delta}[n]$). A point e of $\mathcal{H}om(\Delta[n], E)$ over (a, b) is a singular simplex $e\colon \Delta[n]\to E$ such that $p\circ e=b$ and $e|\dot{\Delta}[n]=a$.

Since $i/p\colon \mathcal{H}om(\Delta[n], E)\to\mathcal{H}om(i, p)$ is a fibration by 4.3, the fibre $F_{a,b}$ over a point (a, b) of $\mathcal{H}om(i, p)$ is a Kan complex. It follows that two vertices of a same connected component of $F_{a,b}$ are the end-points of a same 1-simplex: in other words, consider two singular simplices $e_0\colon \Delta[n]\to E$ and $e_1\colon \Delta[n]\to E$ such that $e_0|\dot{\Delta}[n]=e_1|\dot{\Delta}[n]=a$ and $pe_0=pe_1=b$; the vertices e_0 and e_1 belong to the same connected component of the fibre $F_{a,b}$ if there is a morphism $h\colon \Delta[1]\times\Delta[n]\to E$ such that $b|\Delta[1]\times\dot{\Delta}[n]$ and $p\circ h$ factor through the canonical projuctions of $\Delta[1]\times\dot{\Delta}[n]$ and $\Delta[1]\times\Delta[n]$ onto $\dot{\Delta}[n]$ and $\Delta[n]$. Then we will say simply that e_0 and e_1 are *B-equivalent.* We will also say that the elements of E_n canonically associated with e_0 and e_1 are B-equivalent.

Lemma: Two degenerate simplices of E of dimension n are B-equivalent if and only they if are equal.

This lemma follows directly from 4.3.3 below:

4.3.3. *Lemma: Let x and y be two degenerate simplices of dimension n of a simplicial set E. The relations $d_i x=d_i y$ for $0\leqq i\leqq n$ imply the equality $x=y$.*

Let p and q be such that $x=s_p d_p x$ and $y=s_q d_q x$. If $p=q$, the lemma is proved. If not, suppose that $p<q$. Then we have $x=s_p d_p x=s_p d_p y=s_p d_p s_q d_q y=s_p s_{q-1} d_p d_q y=s_q s_p d_p d_q y=s_q\xi$ with $\xi=s_p d_p d_q y$. We deduce that $d_q x=d_q s_q\xi=\xi$ and $x=s_q d_q x=s_q d_q y=y$.

5. Minimal Fibrations

5.1. Let $p\colon E\to B$ be a fibration and i the inclusion of $\dot{\Delta}[n]$ into $\Delta[n]$. We say that p is a *minimal fibration* (or that E is minimal over B) if two B-equivalent simplices of E (4.3.2) are always equal. It is equivalent to say that each connected component of a fibre of i/p: $\mathcal{H}om(\Delta[n], E)\to\mathcal{H}om(i, p)$ of 4.3.2 has exactly one vertex.

Since i/p is a fibration we can state this last condition in a slightly different way: let t_1 and t_2 be two 1-simplices of $\mathcal{H}om(\Delta[n], E)$ such that $(i/p)t_1=(i/p)t_2$ and $d_e t_1=d_e t_2$ ($e=0$ or 1). Since i/p is a fibration, there is a 2-simplex σ of $\mathcal{H}om(\Delta[n], E)$ such that $d_{1+e}\sigma=t_1$, $d_e\sigma=t_2$ and $(i/p)\sigma=s_e t$ [where $t=(i/p)t_1=(i/p)t_2$]. This shows that $d_{1-e}(t_1)$ and $d_{1-e}(t_2)$ are B-equivalent; hence these vertices coincide when E is minimal over B, and the converse is clear. In other words, we see that p is a minimal fibration if and only if for each diagram (Fig. 92)

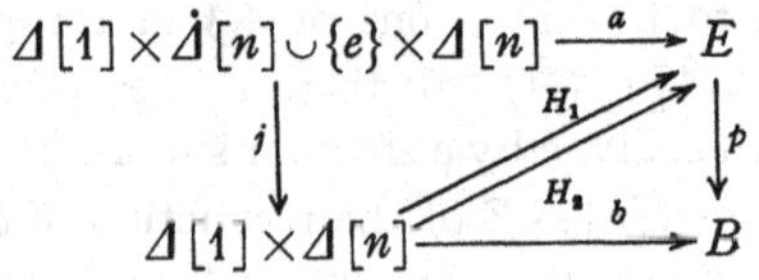

$n\in\mathbb{N}$, $e=0$ or 1

Fig. 92

such that $a=H_1 j=H_2 j$ and $b=pH_1=pH_2$, the restriction of H_1 and H_2 to $\{1-e\}\times\Delta[n]$ coincide.

Note the following two properties of minimal fibrations:

5.1.1. If the commutative square

$$\begin{array}{ccc} E' & \longrightarrow & E \\ {\scriptstyle p'}\downarrow & & \downarrow{\scriptstyle p} \\ B' & \longrightarrow & B \end{array}$$

is cartesian and if p is a minimal fibration, the same holds for p'.

5.1.2. For each commutative diagram (Fig. 93)

$$\begin{array}{ccccc} E & \xrightarrow{u} & E' & \xrightarrow{v} & E \\ {\scriptstyle p}\downarrow & & {\scriptstyle p'}\downarrow & & {\scriptstyle p}\downarrow \\ B & \xrightarrow{u'} & B' & \xrightarrow{v'} & B \end{array}$$

Fig. 93

such that $v\circ u=\mathrm{Id}\,E$ and $v'\circ u'=\mathrm{Id}\,B$, p is a minimal fibration if p' is one (see IV, 3.1.1 iii bis).

5.2. *Theorem* (existence of minimal fibrations): *For each fibration $p\colon E\to B$, there is a simplicial subset E' of E such that the restriction p' of p to E' is a minimal fibration and is a deformation retract of p relatively to B.*

Consider the B-equivalence relation defined in 4.3.2, and choose a simplex in each equivalence class (we will say that such a simplex, as well as the associated singular simplex are *selected*); we require only that each degenerate simplex should be selected, which is possible by lemma 4.3.2.

Let then E' be a subcomplex of E whose simplices are all selected, and which is maximal for this property. Then any selected n-simplex e of E, such that $\tilde{e}|\dot{\Delta}[n]$ factors through E', belongs to E': for let E'' be the smallest subcomplex of E containing E' and e; the simplices of E'' belong to E', are degenerate, or coincide with e; hence they are all selected; hence we have $E''=E'$.

Let us shows now that the restriction p' of p to E' is a deformation retract of p relatively to B, which will imply our theorem, by IV, 3.1.1: let (D, h) be a pair formed by a subcomplex D of E containing E' and by a homotopy $h\colon \Delta[1]\times D\to E$ between the inclusion of D into E and a retraction of D into E', and such that the homotopies $h|\Delta[1]\times E'$ and $p\circ h$ factor through the canonical projections of $\Delta[1]\times E'$ and $\Delta[1]\times D$ onto E' and D. We can obviously suppose that (D, h) is maximal for these properties, and show then that $D=E$:

If this is not true, there is a simplex e of E which does not belong to D, and whose dimension is as small as possible. Then the restriction of the singular simplex $\tilde{e}$ to $\dot{\Delta}[n]$ factors through D, and the smallest subcomplex D' of E which contains D and e make the squares of Fig. 94 cocartesian (II, 3.7).

$$\begin{array}{ccc} \dot{\Delta}[n] & \longrightarrow & D \\ {\scriptstyle\text{incl.}}\downarrow & & \downarrow{\scriptstyle\text{incl.}} \\ \Delta[n] & \xrightarrow{\tilde{e}} & D' \end{array} \qquad \begin{array}{ccc} \Delta[1]\times\dot{\Delta}[n] & \longrightarrow & \Delta[1]\times D \\ \downarrow{\scriptstyle\text{incl.}} & & \downarrow{\scriptstyle\text{incl.}} \\ \Delta[1]\times\Delta[n] & \xrightarrow{\Delta[1]\times\tilde{e}} & \Delta[1]\times D' \end{array}$$

Fig. 94

In order to contradict the maximality of (D, h), by extending h to D', it will then be sufficient to construct a homotopy $k\colon \Delta[1]\times\Delta[n]\to E$, connecting $\tilde{e}$ with a selected singular simplex, whose restriction to $\Delta[1]\times\dot{\Delta}[n]$ coincide with that of $h\circ(\Delta[1]\times\tilde{e})$, and for which $p\circ k$ factors through the canonical projection of $\Delta[1]\times\Delta[n]$ onto $\Delta[n]$. This is equivalent to say that, in the fibration i/p of 4.3.2, we can lift any edge in such a way that the lifting of the origin a is given and that the lifting of the end b is chosen among given representatives of the different connected components of the fibre of b. This is obviously possible.

5.3. *Theorem: Two minimal fibrations $p\colon E\to B$ and $p'\colon E'\to B$ which have the same homotopy type relatively to B,* are *isomorphic relatively to* B (4.2.1 and 4.2).

Let $u\colon E\to E'$ and $v\colon E'\to E$ be morphisms of complexes over B such that vu and $\operatorname{Id} E$ on the one hand, uv and $\operatorname{Id} E'$ on the other, are homotopic relatively to B. We will deduce from lemma 5.3.1 below that uv and vu are isomorphisms relatively to B, so that the same will hold for u and v.

5.3.1. *Lemma: Let $p\colon E\to B$ be a minimal fibration, and let $u\colon E\to E$ be a morphism such that $pu=p$. If u is homotopic to* $\operatorname{Id} E$ *relatively to* B (4.2.1), *u is an isomorphism of $\Delta^\circ\mathscr{E}$.*

By 4.2, $\mathscr{H}om_B(p,p)$ is a Kan complex. It follows that u is homotopic to $\operatorname{Id} E$ relatively to B if and only if there is a homotopy h between $\operatorname{Id} E$ and u such that ph factors through the canonical projection of $\Delta[1]\times E$ onto E.

a) Let us show first that u is a monomorphism, by proving inductively on n, that $u_n\colon E_n\to E_n$ is an injection. We begin the induction with $n=-1$, by defining $E_{-1}=\emptyset$. Suppose then that u_r is injective for $r<n$ and $n\geqq 0$. Let x and x' be two n-simplices such that $u_n(x)=u_n(x')$. The singular simplices $\tilde{x}$ and $\tilde{x}'$ associated with x and x' are such that $\tilde{x}\,|\,\dot{\Delta}[n]=\tilde{x}'\,|\,\dot{\Delta}[n]$ (by induction hypothesis) and $p\circ\tilde{x}=p\circ\tilde{x}'$. Hence we have a commutative diagram (Fig. 95)

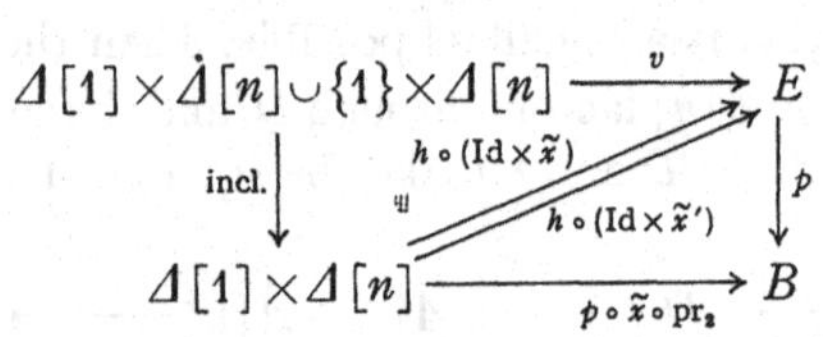

Fig. 95

where v is the common restriction of both $h\circ(\operatorname{Id}\times\tilde{x})$ and $h\circ(\operatorname{Id}\times\tilde{x}')$ to $\Delta[1]\times\dot{\Delta}[n]\cup\{1\}\times\Delta[n]$. By 5.1, the morphisms $h\circ(\operatorname{Id}\times\tilde{x})$ and $h\circ(\operatorname{Id}\times\tilde{x}')$ also have the same restriction to $\{0\}\times\Delta[n]$, which is equivalent to the equality $x=x'$.

b) Suppose now that $u_r\colon E_r\to E_r$ is surjective for $r<n$, and consider an n-simplex x of E. For each integer i, $0\leqq i\leqq n$, there is a $(n-1)$-simplex y_i of E such that $u_{n-1}(y_i)=d_i x$. By a), these y_i are determined in a unique way, and define a morphism $y\colon\dot{\Delta}[n]\to E$ such that $p\circ y=p\circ(\tilde{x}\,|\,\dot{\Delta}[n])$.

By definition of y, the composed morphisms

$$\Delta[1]\times\dot{\Delta}[n]\xrightarrow{\operatorname{Id}\times y}\Delta[1]\times E\xrightarrow{h}E$$

and

$$\{1\}\times\Delta[n]\simeq\Delta[n]\xrightarrow{\tilde{x}}E$$

coincide on $\{1\}\times\Delta[n]$ and define a morphism φ such that the diagram of Fig. 96

$$\begin{array}{ccc} \Delta[1]\times\dot{\Delta}[n]\cup\{1\}\times\Delta[n] & \xrightarrow{\varphi} & E \\ \Big\downarrow{\scriptstyle \text{inclusion}} & & \Big\downarrow{\scriptstyle p} \\ \Delta[1]\times\Delta[n] & \xrightarrow[p\circ\tilde{x}\circ \mathrm{pr}_2]{} & B \end{array}$$

Fig. 96

is commutative. Since the morphism on the left is an anodyne extension, we can complete "commutatively" the diagram by a morphism Φ: $\Delta[1]\times\Delta[n]\to E$. Let then $\tilde{z}$ be the restriction of Φ to $\Delta[n]\simeq\{0\}\times\Delta[n]$ and $z\in E_n$ the simplex associated with $\tilde{z}$. We then have the commutative diagram of Fig. 97

$$\begin{array}{ccc} \Delta[1]\times\dot{\Delta}[n]\cup\{0\}\times\Delta[n] & \longrightarrow & E \\ \Big\downarrow{\scriptstyle \text{incl.}} & \nearrow\!\!\!\nearrow{\scriptstyle \Phi,\ h\circ(\mathrm{Id}\times\tilde{z})} & \Big\downarrow{\scriptstyle p} \\ \Delta[1]\times\Delta[n] & \xrightarrow[p\circ\tilde{x}\circ \mathrm{pr}_2]{} & B \end{array}$$

Fig. 97

By 5.1, Φ and $h\circ(\mathrm{Id}\times\tilde{z})$ have the same restriction to $\{1\}\times\Delta[n]$, whence the equality $x=u_n(z)$.

5.3.2. *Corollary: Two minimal fibrations which are deformation retracts of $p: E\to B$ relatively to B, are isomorphic relatively to B* (4.2).

5.4. Recall (III, 4.1) that a morphism $p: E\to B$ of $\Delta^\circ\mathscr{E}$ is called *locally trivial* if, for each $n\in\mathbb{N}$ and each n-simplex x of B, there is a *cartesian* square of the form

$$\begin{array}{ccc} F\times\Delta[n] & \simeq & E \\ {\scriptstyle \mathrm{pr}_2}\Big\downarrow & & \Big\downarrow{\scriptstyle p} \\ \Delta[n] & \xrightarrow{\tilde{x}} & B \end{array}$$

If b is a vertex of the image of $\tilde{x}$, the complex F is obviously isomorphic to the fibre of p over b. For a locally trivial morphism, this fibre is determined, up to isomorphism, by the connected component of B which contains b.

Theorem: Each minimal fibration is locally trivial.

By 5.1.1, the fibred product $\Delta[n]\underset{\tilde{x},p}{\times}E$ is minimal over $\Delta[n]$ for each simplex x of B. Hence it is sufficient to show that E is trivial over B (III, 4.1) when B is a standard simplex $\Delta[n]$. In that case, let us write f for the identity morphism of $\Delta[n]$, and g: $\Delta[n]\to\Delta[n]$ for the morphism induced by the constant map of $[n]$ onto the element 0 of $[n]$. Since the homotopy $C(v_n^0)$ of IV, 2.1.3 connects f with g, the fibred products $\Delta[n]\underset{f,p}{\times}E$ and $\Delta[n]\underset{g,p}{\times}E$ are isomorphic relatively to

$\Delta[n]$ by 5.4.1 below. But the former is identified with E, while the latter is simply the product $\Delta[n]\times F$, where F denotes the fibre of p over the image of g.

5.4.1. *Proposition: Let $p: E\to B$ be a minimal fibration, and $f, g: A \rightrightarrows B$ two homotopic morphisms. Then the minimal fibrations $A\underset{f,p}{\times}E\to A$ and $A\underset{g,p}{\times}E\to A$ are isomorphic relatively to A.*

We can easily restrict ourselves to the case where we have a homotopy h connecting f with g. Consider then the commutative diagram (Fig. 98)

$$\begin{array}{ccccc}
A\underset{f,p}{\times}E & \xrightarrow{j_0} & (\Delta[1]\times A)\underset{h,p}{\times}E & \xleftarrow{j_1} & A\underset{g,p}{\times}E \\
\downarrow{\scriptstyle q_0} & & \downarrow{\scriptstyle q} & & \downarrow{\scriptstyle q_1} \\
A & \xrightarrow{i_0} & \Delta[1]\times A & \xleftarrow{i_1} & A
\end{array}$$

Fig. 98

where vertical arrows are canonical projections and where i_0, j_0 (resp. i_1, j_1) are induced by the morphism $\Delta(\partial_1^1)\times A: A\simeq\Delta[0]\times A\to\Delta[1]\times A$ (resp. by the morphism $\Delta(\partial_1^0)\times A$). On the other hand, let h_0 (resp. h_1) be the homotopy between $\Delta(\partial_1^1\circ\sigma_0^0)$ and the identity morphism of $\Delta[1]$ (resp. between the identity morphism of $\Delta[1]$ and $\Delta(\partial_1^0\circ\sigma_0^0)$) (see II, 2.1). By lemma 4.4.2 below, there is a retraction r_0 of j_0 (resp. r_1 of j_1) and a homotopy k_0 (resp. k_1) connecting $j_0 r_0$ with the identity morphism of $(\Delta[1]\times A)\underset{h,p}{\times}E$ (resp. connecting the identity morphism of $(\Delta[1]\times A)\underset{h,p}{\times}E$ with $j_1 r_1$), which is "compatible" with the homotopy $h_0\times A$ (resp. $h_1\times A$) of the base $\Delta[1]\times A$. It follows easily that $r_1 j_0$ and $r_0 j_1$ are isomorphisms (inverse to each other) of the category of complexes over A modulo homotopy (4.2.1). Hence q_0 and q_1 have the same homotopy type relatively to A and the proposition follows from 5.3.

5.4.2. *Lemma: Let $p: E\to B$ be a fibration, $j: A\to B$ a monomorphism of $\Delta^\circ\mathscr{E}$, $q: B\to A$ a retraction of j, and h a homotopy connecting* Id B *with jq. Then there is a retraction r of the canonical projection* pr_2: $A\underset{j,p}{\times}E\to E$ *and a homotopy k connecting* Id E *with* $(\mathrm{pr}_2)\, r$*, such that the following square is commutative:*

$$\begin{array}{ccc}
\Delta[1]\times E & \xrightarrow{k} & E \\
\downarrow{\scriptstyle \Delta[1]\times p} & & \downarrow \\
\Delta[1]\times B & \xrightarrow{h} & B
\end{array}$$

(We would obviously have a similar statement if h were a homotopy connecting jq with Id B).

Proof of the lemma: We are looking for an edge k of the complex $\mathscr{H}\!om(E, E)$, whose starting point is the identity morphism of E. The image of k under the morphism $\mathscr{H}\!om(\mathrm{pr}_2, E)$ must be the composition

$$a:\ \Delta[1]\times(A\underset{j,p}{\times}E)\xrightarrow{\text{can.}} A\underset{j,p}{\times}E\xrightarrow{\mathrm{pr}_2} E$$

The image of k under the morphism $\mathcal{H}om(E, p)$ must be the composition

$$b:\ \Delta[1]\times E \xrightarrow{\Delta[1]\times p} \Delta[1]\times B \xrightarrow{\ h\ } B.$$

But (a, b) is an edge of the complex $\mathcal{H}om(\mathrm{pr}_2, p)$ (4.1.1). Since pr_2 is a monomorphism, there is, by 4.3 a lifting of (a, b) into $\mathcal{H}om(E, E)$.

5.4.3. *Corollary: All fibres of a fibration with connected base have the same homotopy type* (i.e. are isomorphic objects of $\overline{\Delta^\circ\mathcal{E}}$).

By 5.2, we are lead back to theorem 5.4.

5.5. Let us return now to the homotopic category $\mathcal{H}$ (resp. to the pointed homotopic category $.\mathcal{H}$) and let us call *representative* of a morphism f of $\mathcal{H}$ (resp. of $.\mathcal{H}$) any morphism φ of $\Delta^\circ\mathcal{E}$ (resp. $.\Delta^\circ\mathcal{E}$) whose image in $\mathcal{H}$ (resp. $.\mathcal{H}$) is isomorphic to f (4.1). Since each morphism f of $\mathcal{H}$ (resp. of $.\mathcal{H}$) is associated with a diagram of $\overline{\Delta^\circ\mathcal{E}}$ (resp. of $.\overline{\Delta^\circ\mathcal{E}}$) of the form

$$\begin{array}{ccc} x & & y \\ & \searrow^{\varphi} & \downarrow \sigma \\ & & y' \end{array}$$

where σ is an anodyne extension (I, 2.3), we see that each morphism has a representative. To complete this paragraph, we will see that *one can always choose as representative a minimal* fibration, i.e. a morphism having remarkabel "geometric" properties (5.4).

5.5.1. *Proposition: For each morphism p: $E\to B$ of $\Delta^\circ\mathcal{E}$, there is a commutative triangle*

$$\begin{array}{ccc} E & \xrightarrow{\ a\ } & E' \\ & {\scriptstyle p}\searrow \quad \swarrow{\scriptstyle p'} & \\ & B & \end{array}$$

such that a is an anodyne extension and p' a fibration.

Let us call p-horn any 4-tuple $\gamma=(n, k, u, v)$ formed by an integer $n\geqq 1$, an integer k such that $0\leqq k\leqq n$, and morphisms u and v of $\Delta^\circ\mathcal{E}$ making the square of Fig. 99 commutative:

$$\begin{array}{ccc} \Lambda^k[n] & \xrightarrow{\ u\ } & E \\ \downarrow{\scriptstyle \text{incl.}} & & \downarrow{\scriptstyle p} \\ \Delta[n] & \xrightarrow{\ v\ } & B \end{array}$$

Fig. 99

Write $n(\gamma)$, $k(\gamma)$, $u(\gamma)$, and $v(\gamma)$ for the components of γ. We then have a commutative square (Fig. 100)

$$\begin{array}{ccc} \coprod_\gamma \Lambda^{k(\gamma)}[n(\gamma)] & \xrightarrow{\ u(p)\ } & E \\ \downarrow{\scriptstyle \text{incl.}} & & \downarrow{\scriptstyle p} \\ \coprod_\gamma \Delta[n(\gamma)] & \xrightarrow{\ v(p)\ } & B \end{array}$$

Fig. 100

where γ runs through the p-horns, and where the components of $u(p)$ [resp. of $v(p)$] are the morphisms $u(\gamma)$ [resp. $v(\gamma)$]. Since the vertical arrow on the left is an anodyne extension, the same holds for the canonical morphism $w(p)$ from E to the amalgamated sum E_1^p of E and $\coprod_\gamma \Delta[n(\gamma)]$ under $\coprod_\gamma \Lambda^{k(\gamma)}[n(\gamma)]$. If $\pi_1(p)\colon E_1^p \to B$ denotes the morphism induced by the square above, and if we define $\pi_{n+1}(p) = \pi_1(\pi_n(p))$ and $E_{n+1}^p = (E_n^p)_1^{\pi_n(p)}$, we get the following commutative diagram of $.\Delta^\circ\mathscr{E}$ (Fig. 101)

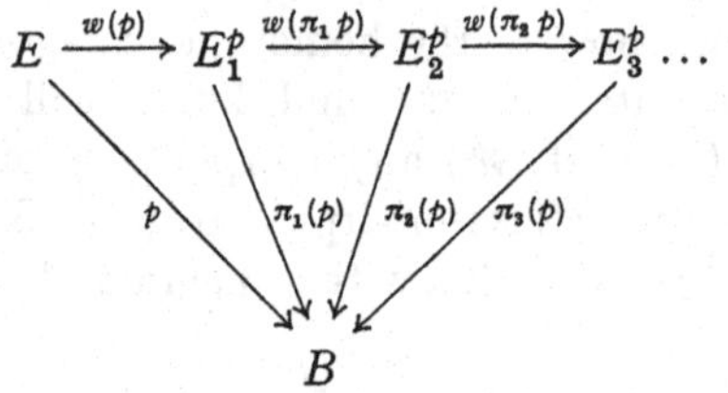

Fig. 101

If E_∞^p is the direct limit of the E_n^p, this diagram induces morphisms $w_\infty(p)\colon E \to E_\infty^p$ and $\pi_\infty(p)\colon E_\infty^p \to B$ of $.\Delta^\circ\mathscr{E}$. Since $w(\pi_i(p))$ is an anodyne extension for all i, $w_\infty(p)$ is an anodyne extension; moreover, it is clear that $\pi_\infty(p)$ is a fibration (see the proof of IV, 3.2).

5.5.2. From 5.5.1 and 5.2, we deduce that, for each morphism $p\colon E \to B$ of $\Delta^\circ\mathscr{E}$, there is a commutative triangle

$$\begin{array}{ccc} E & \xrightarrow{a} & E' \\ & {\scriptstyle p}\searrow \quad \swarrow{\scriptstyle p'} & \\ & B & \end{array}$$

such that p' is a minimal fibration and that the image of a in the homotopic category $\mathscr{H}$ is invertible. This proves our statement of 5.5.

Moreover, if we consider again the proof of 5.2, in the case where E and B have base points e_0 and b_0, and where $p(e_0) = b_0$, we see that the retracting deformation constructed at that time respects base points, provided that e_0 should be "selected". This proves 5.5 for the case of pointed complexes.

5.6. A Kan complex A is said to be *minimal* if the morphism $p\colon A \to \Delta[0]$ is a minimal fibration.

Lemma: Any minimal pointed Kan complex A, such that $\Pi_n A$ reduces to a point for all $n \in \mathbb{N}$, is isomorphic to $\Delta[0]$.

Let a be the base point of A. We obviously have an isomorphism $Sk^{-1}A \simeq Sk^{-1}\Delta[0]$. Suppose then that $Sk^r A$ is isomorphic to $Sk^r \Delta[0]$ for $r < n$, and let us show that A has only one n-simplex, and hence that $Sk^n A$ is isomorphic to $Sk^n \Delta[0]$. If this were not true, A_n would contain a non degenerate simplex x. The singular simplex $\tilde{x}$: $\Delta[n] \rightarrow A$ associated with x sends $\dot{\Delta}[n]$ into $Sk^{n-1}A$, i.e. into the complex $\{a\}$ generated by a. Hence $\tilde{x}$ factors as follows:

$$\Delta[n] \xrightarrow{\text{can.}} \Delta[n]/\dot{\Delta}[n] \xrightarrow{y} A.$$

The canonical bijection

$$\Pi_n A \simeq .\mathscr{H}(\Delta[n]/\dot{\Delta}[n], A)$$

of 3.4 associates with such a y an element of $\Pi_n A$. Since $\Pi_n A$ reduces to a point, y is homotopic to the zero morphism (relatively to the base points of $\Delta[n]/\dot{\Delta}[n]$ and A). In other words, $\tilde{x}$ is $\Delta[0]$-equivalent to the zero morphism in the sense of 4.3.2. Since A is minimal, $\tilde{x}$ is the zero morphism.

5.6.1. *Whitehead's theorem: Let f: $X \rightarrow Y$ be a morphism of the pointed homotopic category, X and Y being connected. Then f is an isomorphism of $.\mathscr{H}$ if and only if $\Pi_n f$ is an isomorphism for each integer $n \geqq 1$.*

The maps $\Pi_n f$: $\Pi_n X \rightarrow \Pi_n Y$ are obviously bijective if f is an isomorphism. In order to prove the converse, we can certainly replace f by any isomorphic morphism of $.\mathscr{H}$. Hence, by 5.5, we can suppose that X and Y are Kan complexes, and that f is a minimal fibration with fibre A. The exact sequence of homotopy groups (3.3)

$$\cdots \rightarrow \Pi_1 A \rightarrow \Pi_1 X \simeq \Pi_1 Y \rightarrow \Pi_0 A \rightarrow \Pi_0 X \simeq \Pi_0 Y$$

shows then that $\Pi_n A$ has only one point for all $n \in \mathbb{N}$. Hence by 5.6, A is isomorphic to $\Delta[0]$ in the category $\Delta^\circ \mathscr{E}$. Since f is supposed to be locally trivial, f is an isomorphism of $\Delta^\circ \mathscr{E}$.

Chapter Seven

Combinatorial Description of Topological Spaces

1. Geometric Realization of the Homotopic Category

1.1. In this chapter, topological spaces will be denoted by thick capital letters $\boldsymbol{X}, \boldsymbol{Y}, \boldsymbol{Z}$ .... We will write $\mathscr{H}om(\boldsymbol{X}, \boldsymbol{Y})$ for the simplicial set whose n-th component is $\mathscr{T}op(\boldsymbol{\Delta}^n \times \boldsymbol{X}, \boldsymbol{Y})$, and $\Pi_0(\boldsymbol{X}, \boldsymbol{Y})$ will be the set of connected components of $\mathscr{H}om(\boldsymbol{X}, \boldsymbol{Y})$. The morphisms of complexes

$$\nu_{\boldsymbol{X},\boldsymbol{Y},\boldsymbol{Z}}: \mathscr{H}om(\boldsymbol{X}, \boldsymbol{Y}) \times \mathscr{H}om(\boldsymbol{Y}, \boldsymbol{Z}) \rightarrow \mathscr{H}om(\boldsymbol{X}, \boldsymbol{Z}),$$

which are defined "as in V, 7.1", induce then, by passing to connected components, maps

$$\varphi_{\boldsymbol{X},\boldsymbol{Y},\boldsymbol{Z}}: \Pi_0(\boldsymbol{X}, \boldsymbol{Y}) \times \Pi_0(\boldsymbol{Y}, \boldsymbol{Z}) \rightarrow \Pi_0(\boldsymbol{X}, \boldsymbol{Z}).$$

These maps are the composition maps of *the category $\overline{\mathcal{Top}}$ of Topological spaces modulo homotopy*: its objects are the topological spaces, its sets of morphisms, the sets $\Pi_0(\boldsymbol{X}, \boldsymbol{Y})$.

1.2. Now let us consider again the pair of adjoint functors

$$S: \mathcal{Top} \to \Delta^\circ\mathcal{E} \quad \text{and} \quad |?|: \Delta^\circ\mathcal{E} \to \mathcal{Top}$$

of III, 1.4. For any complex X, topological space $\boldsymbol{Y}$ and integer n, we then have a natural isomorphism

$$(*) \qquad \mathcal{Top}(|\Delta[n] \times X|, \boldsymbol{Y}) \simeq \Delta^\circ\mathcal{E}(\Delta[n] \times X, S\boldsymbol{Y}).$$

Since $|\Delta[n]|$ is the compact space $\boldsymbol{\Delta}^n$ and since $|\Delta[n] \times X|$ and $\boldsymbol{\Delta}^n \times |X|$ are naturally isomorphic by III, 3.1 and III, 3.1.1, we have natural isomorphisms

$$\mathcal{Top}(\boldsymbol{\Delta}^n \times |X|, \boldsymbol{Y}) \simeq \Delta^\circ\mathcal{E}(\Delta[n] \times X, S\boldsymbol{Y}),$$

hence an isomorphism

$$(**) \qquad \mathcal{Hom}(|X|, \boldsymbol{Y}) \simeq \mathcal{Hom}(X, S\boldsymbol{Y})$$

and finally by passing to connected components, an isomorphism

$$\overline{\mathcal{Top}}(|X|, \boldsymbol{Y}) \simeq \overline{\Delta^\circ\mathcal{E}}(X, S\boldsymbol{Y}).$$

As in IV, 1.5, we deduce from these "equalities" that *the geometric realization functor defines, by passing to the quotient, a functor from $\overline{\Delta^\circ\mathcal{E}}$ to $\overline{\mathcal{Top}}$* (which will also be written $|?|$). Similarly, S *defines by passing to the quotient a functor which is right adjoint to* $|?|: \overline{\Delta^\circ\mathcal{E}} \to \overline{\mathcal{Top}}$ (we will also write S for this adjoint functor).

1.3. Let j be an anodyne extension of $\overline{\Delta^\circ\mathcal{E}}$. We will see in 1.6 and 1.7 that the geometric realization $|j|$ of j is an isomorphism of $\overline{\mathcal{Top}}$. Hence the functor $|?|: \overline{\Delta^\circ\mathcal{E}} \to \overline{\mathcal{Top}}$ is the composition of the canonical functor $P_{\bar{A}}: \overline{\Delta^\circ\mathcal{E}} \to \mathcal{H}$ defined in IV, 2.3.1 with a functor

$$\|?\|: \mathcal{H} \to \overline{\mathcal{Top}}$$

which will still be called geometric realization. Moreover, by I, 1.3.1, the functor $\|?\|$ is left adjoint to the composed functor $P_{\bar{A}} \circ S$, which we will note $\mathcal{S}$, and which we will also call singular complex functor.

Fundamental Theorem (MILNOR): *The geometric realization functor*

$$\|?\|: \mathcal{H} \to \overline{\mathcal{Top}}$$

is fully faithful.

The proof of this theorem will be given in paragraph 3. Note however that, by proposition I, 1.3, the fundamental theorem implies that $\mathcal{S}$ induces an equivalence from a category of fractions of $\overline{\mathcal{Top}}$ onto $\mathcal{H}$. Taking IV, 3.2.1 into account, we see finally that *the homotopic category*

$\mathcal{H}$ is equivalent to the category of Kan complexes modulo homotopy; or to the full subcategory of $\overline{\mathcal{Top}}$ formed all topological spaces which are isomorphic in $\overline{\mathcal{Top}}$ to the geometric realization of a simplicial set; or finally, to a category of fractions of $\overline{\mathcal{Top}}$.

1.4. We will show now that the geometric realization of an anodyne extension of $\Delta^\circ\mathcal{E}$ is an invertible morphism of $\overline{\mathcal{Top}}$: in order to do this, recall first that a continuous map $\boldsymbol{p}: \boldsymbol{X} \to \boldsymbol{B}$ is called a *fibration in the sense of* SERRE if, for each integer $n \geq 1$ and each pair of continuous maps

$$\boldsymbol{f}: \boldsymbol{I}^n \to \boldsymbol{B},\ \boldsymbol{g}: \boldsymbol{I}^{n-1} \times \{0\} \to \boldsymbol{X}$$

such that $\boldsymbol{p} \circ \boldsymbol{g} = \boldsymbol{f} | \boldsymbol{I}^{n-1} \times \{0\}$, there is a continuous map $\boldsymbol{h}: \boldsymbol{I}^n \to \boldsymbol{X}$ such that $\boldsymbol{g} = \boldsymbol{h} | \boldsymbol{I}^{n-1} \times \{0\}$ and $\boldsymbol{f} = \boldsymbol{p} \circ \boldsymbol{h}$ ($\boldsymbol{I}$ denotes the interval $[0, 1]$).

The notion of a fibration in the sense of SERRE has a local character with respect to the base: if $\boldsymbol{p}: \boldsymbol{X} \to \boldsymbol{B}$ is a continuous map and if each point of $\boldsymbol{B}$ has an open neighborhood $\boldsymbol{U}$ such that the map from $\boldsymbol{p}^{-1}(\boldsymbol{U})$ to $\boldsymbol{U}$, which is induced by $\boldsymbol{p}$, is a fibration in the sense of SERRE, then $\boldsymbol{p}$ is a fibration in the sense of SERRE. Indeed, if $\boldsymbol{f}$ and $\boldsymbol{g}$ are the above maps, we can subdivide $\boldsymbol{I}^n$ in cubes small enough so that $\boldsymbol{f}$ maps each cube of the subdivision into an open set $\boldsymbol{U}$ of $\boldsymbol{B}$ such that $\boldsymbol{p}^{-1}(\boldsymbol{U})$ is a fibration over $\boldsymbol{U}$. We can then define gradually, on the cubes of the subdivision, an extension $\boldsymbol{h}$ of $\boldsymbol{g}$.

For instance, if $\boldsymbol{X}$ and $\boldsymbol{B}$ are Kelley spaces (I, 1.5.3) and if $\boldsymbol{p}$ is a trivial morphism in the sense of III, 4.1, then $\boldsymbol{p}$ is obviously a fibration in the sense of SERRE. By above, the conclusion still holds when $\boldsymbol{p}$ is a locally trivial morphism of Kelley spaces. In particular, *if $q: X \to B$ is a locally trivial morphism of $\Delta^\circ\mathcal{E}$, then $|q|: |X| \to |B|$ is a fibration in the sense of* SERRE (VI, 5.4 and II, 4.2).

1.5. *The image of a fibration in the sense of* SERRE *under the singular complex functor is a fibration in the sense of* KAN: let $\boldsymbol{p}: \boldsymbol{X} \to \boldsymbol{B}$ be a fibration in the sense of SERRE. Since S is right adjoint to the geometric realization functor, the existence of a square of $\Delta^\circ\mathcal{E}$ of type (1) is equivalent to the existence of a square of $\mathcal{Top}$ of type (2) (Fig. 102)

$$(1)\quad \begin{array}{ccc} \Delta^k[n] & \xrightarrow{e} & S\boldsymbol{X} \\ {\scriptstyle \text{incl.}}\downarrow & & \downarrow{\scriptstyle S\boldsymbol{p}} \\ \Delta[n] & \xrightarrow{x} & S\boldsymbol{B} \end{array} \qquad\qquad (2)\quad \begin{array}{ccc} |\Delta^k[n]| & \xrightarrow{\boldsymbol{f}} & \boldsymbol{X} \\ {\scriptstyle |\text{incl.}|}\downarrow & & \downarrow{\scriptstyle \boldsymbol{p}} \\ |\Delta[n]| & \xrightarrow{\boldsymbol{y}} & B \end{array}$$

Fig. 102

Since $\boldsymbol{p}$ is a fibration in the sense of SERRE, and since the continuous map $|\text{incl.}|$ is obviously isomorphic to the inclusion of $\boldsymbol{I}^{n-1} \times \{0\}$ into $\boldsymbol{I}^n$ (VI, 4.1), there is a continuous map $\boldsymbol{s}: |\Delta[n]| \to X$ such that $\boldsymbol{p} \circ \boldsymbol{s} = \boldsymbol{y}$ and $\boldsymbol{s} \circ |\text{incl.}| = \boldsymbol{f}$. The morphism $r: \Delta[n] \to S\boldsymbol{X}$, canonically associated with $\boldsymbol{s}$, is then such that $r | \Delta^k[n] = e$ and $x = (S\boldsymbol{p}) r$.

1.6. When $\boldsymbol{B}$ reduces to a point, the unique map $\boldsymbol{p}$ from $\boldsymbol{X}$ to $\boldsymbol{B}$ is obviously a fibration in the sense of SERRE. Since $S\boldsymbol{B}$ is then identified with $\Delta[0]$, it follows that *the singular complex* $S\boldsymbol{X}$ *of any topological space* $\boldsymbol{X}$ *is a Kan complex.*

In order to show that the geometric realization $|j|$ of an anodyne extension j of $\overline{\Delta^\circ\mathcal{E}}$ is an invertible morphism of $\overline{\mathcal{T}op}$, it is thus sufficient to apply the results of IV, 3.1.5 and the equalities $\overline{\mathcal{T}op}(|j|, \boldsymbol{X}) \simeq \overline{\Delta^\circ\mathcal{E}}(j, S\boldsymbol{X})$.

1.7. To conclude, we will give a geometric demonstration of the fact that the geometric realization of an anodyne extension of $\overline{\Delta^\circ\mathcal{E}}$ is an invertible morphism of $\overline{\mathcal{T}op}$. To this end, consider the subset E of $\mathfrak{Ar}\,\mathcal{T}op$ determined by the homeomorphisms onto subspaces satisfying the following property: if $\boldsymbol{i}\colon \boldsymbol{A}\to\boldsymbol{B}$ is an injection belonging to E, there is a continuous map $\boldsymbol{h}\colon [\alpha,\beta]\times\boldsymbol{B}\to\boldsymbol{B}$ ($\alpha,\beta\in\boldsymbol{R}$) such that $\boldsymbol{h}(\beta,\boldsymbol{b})=\boldsymbol{b}$, $\boldsymbol{h}(t,\boldsymbol{i}(\boldsymbol{a}))=\boldsymbol{i}(\boldsymbol{a})$, $\boldsymbol{h}(\alpha,\boldsymbol{b})\in\boldsymbol{i}(\boldsymbol{A})$ for any $\boldsymbol{b}$ in $\boldsymbol{B}$, $\boldsymbol{a}$ in $\boldsymbol{A}$, and t in $[\alpha,\beta]$ (if $\boldsymbol{i}$ is the inclusion of the subspace $\boldsymbol{A}$ into $\boldsymbol{B}$, we say that $\boldsymbol{A}$ is a deformation retract of $\boldsymbol{B}$). It is clear that if $\boldsymbol{i}$ belongs to E, it is invertible in $\overline{\mathcal{T}op}$. We will show that if s is an anodyne extension of $\Delta^\circ\mathcal{E}$, then $|s|$ is an element of E. The property we want to prove will then be an immediate consequence of 1.2.

The set E has the properties stated in IV, 2.1, namely:

isomorphisms belong to E.
E is stable under push-out
E is stable under retractions
E is stable under countable compositions and arbitrary direct sums.

Let us verify for instance that E is stable under "countable inclusion". Let

$$\boldsymbol{X}_0\subset\boldsymbol{X}_1\subset\dots\subset\boldsymbol{X}_n\subset\dots\subset\boldsymbol{X}=\varinjlim \boldsymbol{X}_n$$

be a sequence of topological spaces. We suppose that for each n ,we have a retracting deformation $\boldsymbol{h}_n\colon \left[1-\frac{1}{n}, 1-\frac{1}{n+1}\right]\times\boldsymbol{X}_n\to\boldsymbol{X}_n$, and we define $\boldsymbol{j}_n(\boldsymbol{x})=\boldsymbol{h}_n\left(1-\frac{1}{n},\boldsymbol{x}\right)$ (then $\boldsymbol{j}_n(\boldsymbol{x})\in\boldsymbol{X}_{n-1}$). Finally, let $\boldsymbol{H}_n\colon [0,1]\times\boldsymbol{X}_n\to\boldsymbol{X}_n$ be the function defined by

$$\boldsymbol{H}_n(t,\boldsymbol{x})=\begin{cases}\boldsymbol{x} & \text{for each } t>1-\dfrac{1}{n+1}\\ \boldsymbol{h}_{n-p}(t,\boldsymbol{j}_{n-p+1}\circ\dots\circ\boldsymbol{j}_n(\boldsymbol{x})) & \text{for } t\in\left[1-\dfrac{1}{n-p}, 1-\dfrac{1}{n-p+1}\right].\end{cases}$$

The $\boldsymbol{H}_n$ define a continuous map $\boldsymbol{H}\colon [0,1]\times\boldsymbol{X}\to\boldsymbol{X}$ which is a retracting deformation of $\boldsymbol{X}$ onto $\boldsymbol{X}_0$.

On the other hand, since the geometric realization of the inclusion $\Lambda^k[n]\to\Delta[n]$ belongs to E, it follows from IV, 2.1.4 that E contains the geometric realization of all anodyne extensions of $\Delta^\circ\mathcal{E}$. Q.E.D.

2. Geometric Realization of the Pointed Homotopic Category

2.1. Let $(\boldsymbol{X}, \boldsymbol{x}_0)$ be a pointed topological space. The canonical map $\boldsymbol{x}_0'$ from Δ° to X, whose image is $\boldsymbol{x}_0$, is then a vertex of the singular complex $S\boldsymbol{X}$. We will define then

$$.S(\boldsymbol{X}, \boldsymbol{x}_0) = (S\boldsymbol{X}, \boldsymbol{x}_0')$$

and we will also call $.S\colon .\mathcal{T}op \to .\Delta^\circ\mathcal{E}$ the singular complex functor. Similarly, let (T, t_0) be a pointed simplicial set. With the singular complex $\tilde{t}_0\colon \Delta[0] \to T$ is then associated a map $|\tilde{t}_0|\colon \Delta^\circ \to |T|$ whose image will be denoted by $\boldsymbol{t}_0'$. Again we will define

$$.|T, T_0| = (|T|, \boldsymbol{t}_0')$$

and we will say that $.|?|\colon .\Delta^\circ\mathcal{E} \to .\mathcal{T}op$ is the geometric realization functor.

It is clear that the bijection

$$\mathcal{T}op(|T|, \boldsymbol{X}) \simeq \Delta^\circ\mathcal{E}(T, S\boldsymbol{X})$$

induces a bijection from the subset $.\mathcal{T}op(.|T|, \boldsymbol{X})$ of $\mathcal{T}op(|T|, \boldsymbol{X})$ onto the subset $.\Delta^\circ\mathcal{E}(T, .S\boldsymbol{X})$ of $\Delta^\circ\mathcal{E}(T, S\boldsymbol{X})$. Hence $.|?|$ *is left adjoint to* $.S$.

2.2. Now let $\boldsymbol{Z}$ and $\boldsymbol{Y}$ be two pointed topological spaces, with base points $\boldsymbol{z}_0$ and $\boldsymbol{y}_0$. We will denote then by $\mathcal{H}om.(\boldsymbol{Z}, \boldsymbol{Y})$ the subcomplex of the complex $\mathcal{H}om(\boldsymbol{Z}, \boldsymbol{Y})$ of 1.1, whose n-th component is formed by all continuous maps $\boldsymbol{f}$ from $\Delta^n \times \boldsymbol{Z}$ to $\boldsymbol{Y}$ which send $\Delta^n \times \{\boldsymbol{z}_0\}$ to $\{\boldsymbol{y}_0\}$ (see V, 7.1). For each pointed complex (X, x_0), the isomorphism (**) of 1.2 induces then an isomorphism

$$\mathcal{H}om.(.|X|, \boldsymbol{Y}) \simeq \mathcal{H}om.(X, .S\boldsymbol{Y}),$$

and hence, by passing to connected components, an isomorphism

$$|.|X|, \boldsymbol{Y}| \simeq |X, .S\boldsymbol{Y}|$$

(Notations are those of chapter V; the set $|X, .S\boldsymbol{Y}|$ has also been denoted by $\Pi_0^{\cdot}(X, .S\boldsymbol{Y})$ in IV, 4.2). Hence we see that the geometric realization functor defines, by passing to the quotient, a functor from $.\overline{\Delta^\circ\mathcal{E}}$ to the category. $\overline{\mathcal{T}op}$ of pointed topological spaces modulo homotopy (this category was denoted by $.\overline{\mathfrak{Top}}$ in V, 7). The functor thus defined will also be denoted by $\underline{.|?|}$. Similarly, $.S$ defines a functor which is right-adjoint to $.|?|\colon .\overline{\Delta^\circ\mathcal{E}} \to .\overline{\mathcal{T}op}$; we will still write $.S$ for this functor.

2.3. For each pointed topological space $\boldsymbol{Y}$, $.S\boldsymbol{Y}$ is a pointed Kan complex by 1.6. Hence the pointed complex $.S\boldsymbol{Y}$ is left closed relatively to the set $.\bar{A}$ of anodyne extensions of $.\overline{\Delta^\circ\mathscr{E}}$ (IV, 4.3). Hence we see as in 1.6, that the geometric realization of an anodyne extension of $.\overline{\Delta^\circ\mathscr{E}}$ is an invertible morphism of $.\overline{\mathscr{Top}}$. Hence $.|?|$ is the composition of the canonical functor

$P_{.\bar{A}}: .\overline{\Delta^\circ\mathscr{E}} \to \mathscr{H}$ with a functor

$$.\|?\|: .\mathscr{H} \to .\overline{\mathscr{Top}}$$

which we will also call geometric realization. By I, 1.3.1, this functor is left adjoint to the "singular complex" functor $P_{.\bar{A}} \circ .S$, which will also be denoted by $.\mathscr{S}$.

Fundamental Theorem (MILNOR): *The geometric realization functor is*

$$.\|?\|: .\mathscr{H} \to .\overline{\mathscr{Top}}$$

is fully faithful.

The demonstration will be given in paragraph 3. We deduce, as usual, that the pointed homotopic category $.\mathscr{H}$ is equivalent to a category of fractions of $.\overline{\mathscr{Top}}$ ....

2.4. Let (Z, z_0) and (T, t_0) be two pointed complexes such that the geometric realization $|T|$ of T is locally compact. By III, 2.1.1 and III, 3.1, the canonical map from $|Z \times T|$ to the cartesian product of the topological spaces $|Z|$ and $|T|$ is a homeomorphism. Moreover, since the geometric realization functor commutes with direct limits, the image of the square of Fig. 103

$$\begin{array}{ccc} Z \times \{t_0\} \cup \{z_0\} \times T & \longrightarrow & Z \times T \\ \downarrow & & \downarrow \\ \Delta[0] & \longrightarrow & Z \wedge T \end{array}$$

Fig. 103

under the geometric realization functor is still a cocartesian square. It follows that $|Z \wedge T|$ *is identified with* $|Z| \wedge |T|$ *when* $|T|$ *is locally compact.*

If we consider now a morphism $f: X \to Y$ of pointed complexes, the geometric realization of the square

$$\begin{array}{ccc} X & \xrightarrow{f} & Y \\ {\scriptstyle \varepsilon}\downarrow & & \downarrow{\scriptstyle if} \\ \Delta[1] \wedge X & \longrightarrow & Cf \end{array}$$

of VI, 2.1 is cocartesian. Since $|\Delta[1] \wedge X|$ is identified with $\boldsymbol{I} \wedge |X|$ by above, we see that $|Cf|$ is simply $C|f|$ (V, 7.3). More generally, the infinite sequence

$$|X| \xrightarrow{|f|} |Y| \xrightarrow{|if|} |Cf| \xrightarrow{|jf|} |\Sigma X| \xrightarrow{|\Sigma f|} |\Sigma Y| \ldots.$$

is identified with the sequence

$$|X| \xrightarrow{|f|} |Y| \xrightarrow{i|f|} C|f| \xrightarrow{j|f|} \Sigma|X| \xrightarrow{\Sigma|f|} \Sigma|Y| \ldots.$$

2.5. The singular complex functor $.\mathscr{S}$ allows us to define the homotopy groups of a pointed topological space $\boldsymbol{T}$ by means of the formula

$$\Pi_n \boldsymbol{T} = \Pi_n(.\mathscr{S}\boldsymbol{T}).$$

By VI, 3.2, we then have ($\boldsymbol{S}^n$ being the n-sphere)

$$\Pi_n \boldsymbol{T} = .\mathscr{H}(\Sigma^n \dot{\Delta}[1], .\mathscr{S}\boldsymbol{T}) \simeq |.\|\Sigma^n \dot{\Delta}[1]\|, \boldsymbol{T}| \simeq |\Sigma^n \boldsymbol{S}^0, \boldsymbol{T}| = |\boldsymbol{S}^n, \boldsymbol{T}|.$$

Hence, by definition, $\Pi_n \boldsymbol{T}$ is identified with the set of morphisms from the sphere of dimension n to $\boldsymbol{T}$, these morphisms being those of the category of pointed topological spaces modulo homotopy.

When $\mathbf{p}$: $\boldsymbol{X} \to \boldsymbol{B}$ *is a morphism of pointed topological spaces, such that* $\mathbf{p}$ *is a fibration in the sense of* Serre (1.4), *and when* $\boldsymbol{F}$ *denotes the fibre of* $\mathbf{p}$ *over the base point of* $\boldsymbol{B}$, it follows from 1.5 that *the sequence*

$$.S\boldsymbol{F} \xrightarrow{\text{incl.}} .S\boldsymbol{X} \xrightarrow{.S\mathbf{p}} .S\boldsymbol{B}$$

is a fibration (VI, 1.5). Hence, by VI, 3.3, we have an exact sequence of homotopy groups

$$\ldots \Pi_2 \boldsymbol{B} \to \Pi_1 \boldsymbol{F} \to \Pi_1 \boldsymbol{X} \xrightarrow{\Pi_1 \mathbf{p}} \Pi_1 \boldsymbol{B} \to \Pi_0 \boldsymbol{F} \to \Pi_0 \boldsymbol{X} \xrightarrow{\Pi_0 \mathbf{p}} \Pi_0 \boldsymbol{B}.$$

3. Proof of Milnor's Theorem

3.1. We know that with the adjunction isomorphisms

$$\overline{\mathscr{T}op}(\|X\|, \boldsymbol{Y}) \simeq \mathscr{H}(X, \mathscr{S}\boldsymbol{Y}) \quad \text{and} \quad .\overline{\mathscr{T}op}(.\|X\|, \boldsymbol{Y}) \simeq .\mathscr{H}(X, .\mathscr{S}\boldsymbol{Y}),$$

which we described in 1.3 and 2.3, are canonically associated functor morphisms

$$\Psi X \colon X \to \mathscr{S}\|X\| \quad \text{and} \quad .\Psi X \colon X \to .\mathscr{S}.\|X\|.$$

By I, 1.3, theorems 1.3 and 2.3 will be proved if we show that the morphisms ΨX (resp. $.\Psi X$) are isomorphisms of $\mathscr{H}$ (resp. of $.\mathscr{H}$) when X runs through the objects of $\mathscr{H}$ (resp. of $.\mathscr{H}$). This will be a consequence of the following lemma:

Lemma (Milnor): *For each object* X *of* $.\mathscr{H}$, *and for each integer* $n \geqq 0$, *the maps*

$$\Pi_n(.\Psi X) \colon \Pi_n X \to \Pi_n(.\mathscr{S}.\|X\|)$$

are bijections.

The lemma is proved by induction on n. It is clear that $\Pi_0(.\Psi X)$ is an isomorphism. On the other hand, since each object X of $.\mathscr{H}$ is isomorphic (in $.\mathscr{H}$) to a pointed Kan complex, we can suppose that X is a Kan complex. In that case, the morphism

$$\mathscr{H}om.(\Delta[1], X) \to \mathscr{H}om.(\dot{\Delta}[1], X) \simeq X$$

of $.\Delta^\circ\mathscr{E}$, which is induced by the inclusion of $\dot{\Delta}[1]$ into $\Delta[1]$, is a fibration with fibre $\Omega X = \mathscr{H}om.(\Omega, X)$ (VI, 1.2.1).

By VI, 5.2, there is a fibration $F \xrightarrow{i'} E \xrightarrow{p'} X$, which is minimal and which is a deformation retract of p relatively to X. Since $\mathscr{H}om.(\Delta[1], X)$ is contractible by VI, 1.4 (proof of the lemma), E is also contractible. Moreover, p' is locally trivial by VI, 5.4 and

$$.S.|F| \xrightarrow{.S.|i'|} .S.|E| \xrightarrow{.S.|p'|} .S.|X|$$

is a fibration by 1.4 and 1.5. It follows that the commutative diagram of Fig. 104

$$\begin{array}{ccccc} F & \xrightarrow{i'} & E & \xrightarrow{p'} & X \\ {\scriptstyle .\Psi F}\uparrow & & {\scriptstyle .\Psi E}\downarrow & & {\scriptstyle .\Psi X}\downarrow \\ .S.|F| & \to & .S.|E| & \to & .S.|X| \end{array}$$

Fig. 104

induces a morphism between the exact sequences of homotopy groups (Fig. 105).

$$\begin{array}{ccccccc} \ldots \Pi_n E & \longrightarrow & \Pi_n X & \longrightarrow & \Pi_{n-1} F & \longrightarrow & \Pi_{n-1} E \ldots \\ {\scriptstyle \Pi_n(.\Psi E)}\downarrow & & {\scriptstyle \Pi_n(.\Psi X)}\downarrow & & {\scriptstyle \Pi_{n-1}(.\Psi F)}\downarrow & & {\scriptstyle \Pi_{n-1}(.\Psi E)}\downarrow \\ \ldots \Pi_n(.S.|E|) & \longrightarrow & \Pi_n(.S.|X|) & \longrightarrow & \Pi_{n-1}(.S.|F|) & \longrightarrow & \Pi_{n-1}(.S.|E|) \ldots \end{array}$$

Fig. 105

Since E is isomorphic to $\Delta[0]$ in $.\mathscr{H}$, the same holds for $.S.|E|$, so that $\Pi_n E, \Pi_{n-1} E, \Pi_n(.S.|E|)$ and $\Pi_{n-1}(.S.|E|)$ reduce to a point. Hence we have $\Pi_n X \simeq \Pi_{n-1} F$ and $\Pi_n(.S.|X|) \simeq \Pi_{n-1}(.S.|F|)$. Hence we have $\Pi_n X \simeq \Pi_n(.S.|X|)$, by induction on n (for $n=1$, the above isomorphisms come from V, 5.3.3).

3.2. It remains to be shown that ΨX or $.\Psi X$ are invertible:

a) Suppose first that X *is a conneced pointed complex*. Then the fact that $.\Psi X$ is invertible following from 3.1 and from WHITEHEAD's theorem (VI, 5.6.1).

b) Suppose now that X is a *complex without a base point* (i.e. an object of $\mathscr{H}$). It is easy to see that we can restrict ourselves to the case where X is connected, and we choose an arbitrary vertex x_0 of X. Define $\tilde{X} = (X, x_0)$ and write i: $(T, t_0) \rightsquigarrow T$ for the "forgetful" functor from $.\mathscr{H}$ to $\mathscr{H}$. We then have equalities $i(.\mathscr{S}.\|\tilde{X}\|) = \mathscr{S}\|X\|$ and $\Psi X = i(.\Psi\tilde{X})$. Since $.\Psi\tilde{X}$ is invertible by a), ΨX is also invertible.

c) Finally, suppose that X is *an arbitrary pointed complex.* If we apply a) to the connected component of X which contains the base point, and b) to the complex without base point iX, the invertible of $.\Psi X$ will then follow from the following lemma:

Lemma: Let X and X' be two objects of $.\mathscr{H}$, Y and Y' the connected components of X and X' containing the base points. Finally, let $f: X \to X'$ be a morphism of $.\mathscr{H}$ such that if is invertible, as well as the morphism $g: Y \to Y'$ induced by f. Then f is invertible.

The proof of this lemma is left to the reader. It follows obviously from the fact that f is determined by the restriction of f to the connected component of the base point of X on the one hand, and by the restriction of f to the other connected components on the other hand.

Appendix One

Coverings

1. Coverings of a Groupoid

1.1. Let G be a groupoid and R/G the full subcategory of $\mathscr{G}r/G$ whose objects are the morphisms $p: R \to G$ such that for each commutative square

$$\begin{array}{ccc} Sc[0] & \xrightarrow{i} & R \\ {\scriptstyle q}\downarrow & & \downarrow{\scriptstyle p} \\ Sc[1] & \xrightarrow{j} & G \end{array}$$

there is a *unique* morphism $s: Sc[1] \to R$ satisfying the equalities $ps = j$ and $sq = i$. An object of R/G is called a *covering of* G. Recall that, if $p: R \to G$ and $p': R' \to G$ are two coverings of G, a morphism from the first to the second is a morphism $f: R \to R'$ such that $p'f = p$.

1.2. Let $G^\circ\mathscr{E}$ be the category of contravariant functors from the groupoid G to the category of sets. Such a functor is called *a local system* on G (we choose contravariant functors for reasons of notations only). We will first give *an equivalence between the categories* R/G *and* $G^\circ\mathscr{E}$.

Let $p: R \to G$ be a covering of G, and x an object of G. We write $\mathsf{L}(p)(x)$ or $p^{-1}(x)$ for the set of objects of R which project onto x. Let x' be such an object: by definition of coverings, each morphism $\alpha: y \to x$ of G can be lifted in a unique way to a morphism $\alpha': y' \to x'$ of R. Hence the morphism α defines a morphism $L(p)(\alpha): x' \rightsquigarrow y'$ from $\mathsf{L}(p)(x)$ to $\mathsf{L}(p)(y)$. We can check directly that the maps $x \rightsquigarrow \mathsf{L}(p)(x)$ and $\alpha \rightsquigarrow \mathsf{L}(p)(\alpha)$ define a local system $\mathsf{L}(p)$ on G. Write $\mathsf{L}: \mathsf{R}/G \to G^\circ\mathscr{E}$ for the functor $p \rightsquigarrow \mathsf{L}(p)$.

Conversely, let L be a local system, and write $\mathsf{R}(L): R(L) \to G$ for the following groupoid over G: The set of objects of $R(L)$ is the direct sum $\coprod_{x \in G} L(x)$ and $\mathsf{R}(L)$ maps the direct summand $L(x)$ of $\coprod_{x \in G} L(x)$ onto x; the morphisms of $R(L)$ are the pairs (α, x') formed by a morphism $\alpha: y \to x$ of G and an element x' of $L(x)$; The range of (α, x') is $x' \in L(x)$, and its domain is $L(\alpha)(x') \in L(y)$; the image of (α, x') under $\mathsf{R}(L)$ is α; finally, composition is defined by the formula

$$(\alpha, x') \circ (\beta, L(\alpha)(x')) = (\alpha \circ \beta, x').$$

We can check directly that $\mathsf{R}(L)$ is a covering of G. Moreover, if $\mathsf{R}: G^\circ \mathscr{E} \to \mathsf{R}/G$ is the functor $L \rightsquigarrow \mathsf{R}(L)$, we can deduce easily the following theorem:

Theorem: The functors $\mathsf{L}: \mathsf{R}/G \to G^\circ \mathscr{E}$ *and* $\mathsf{R}: G^\circ \mathscr{E} \to \mathsf{R}/G$ *are equivalences quasi inverse to each other.*

1.3. *Corollary: Let* $p: R \to G$ *be a covering of the groupoid* G. *Then the following statements are equivalent.*

(i) *The local system* $\mathsf{L}(p): G^\circ \to \mathscr{E}$ *is a representable functor.*

(ii) R *is simply connected.*

(iii) *The covering* p *is universal; i.e. for each covering* $q: S \to G$, *for each object* x *of* R *and each object* y *of* S *such that* $p(x) = q(y)$, *there is one and only one functor* $f: R \to S$ *such that* $q \circ f = p$ *and* $y = f(x)$.

1.4. Let $f: H \to G$ be a morphism of groupoids. For each covering $p: R \to G$, we write $f^{-1}(p)$ for the projection of the fibred product $H \underset{G}{\times} R$ onto the first factor. It is clear that $f^{-1}(p)$ is a covering of H; and we write R/f for the functor $p \rightsquigarrow f^{-1}(p)$. Similarly, we write $f^\circ \mathscr{E}: G^\circ \mathscr{E} \to H^\circ \mathscr{E}$ for the restriction $L \rightsquigarrow L \circ f^\circ$. We then have a diagram (Fig. 106)

$$\begin{array}{ccc} \mathsf{R}/H & \xleftarrow{\mathsf{R}/f} & \mathsf{R}/G \\ {\scriptstyle \mathsf{L}}\downarrow\uparrow{\scriptstyle \mathsf{R}} & & {\scriptstyle \mathsf{L}}\downarrow\uparrow{\scriptstyle \mathsf{R}} \\ H^\circ \mathscr{E} & \xleftarrow{f^\circ \mathscr{E}} & G^\circ \mathscr{E} \end{array}$$

Fig. 106

which is commutative up to isomorphisms; in other words, $(f^\circ \mathscr{E}) \circ \mathsf{L}$ is isomorphic to $\mathsf{L} \circ (\mathsf{R}/f)$ and $(\mathsf{R}/f) \circ \mathsf{R}$ is isomorphic to $\mathsf{R} \circ (f^\circ E)$.

Proposition: R/f *is an equivalence if and only* f *is an equivalence.*

If f is an equivalence, $f^\circ \mathscr{E}$ is an equivalence. Conversely, if $f^\circ \mathscr{E}$ is an equivalence, the functor f^*, which is left adjoint to $f^\circ \mathscr{E}$, is an equivalence, and hence it is fully faithful. Since the square

$$\begin{array}{ccc} H^\circ \mathscr{E} & \xrightarrow{f^*} & G^\circ \mathscr{E} \\ {\scriptstyle h^H}\uparrow & & \uparrow{\scriptstyle h^G} \\ H & \xrightarrow{f} & G \end{array}$$

is commutative up to isomorphism (see II, 1.3), and since h^H and h^G are fully faithful, f is fully faithful. Hence f induces an equivalence f' from H *onto* the direct sum G' of all connected components of G which intersect the image of f (by II, 6.1.5). The statement follows then from the fact that the inclusion of G' into G induces an eqiuvalence from $G'^{\circ}\mathscr{E}$ onto $G^{\circ}\mathscr{E}$ if and only if G' is equal to G.

2. Coverings of Groupoids and Simplicial Coverings

2.1. Let X be a simplicial set. We write R/X for the full subcategory of $\Delta^{\circ}\mathscr{E}/X$ whose objects are the morphisms $p\colon E\to X$ such that for each commutative diagram

$$\begin{array}{ccc} \Delta[0] & \xrightarrow{u} & E \\ {\scriptstyle i}\big\downarrow & & \big\downarrow{\scriptstyle p} \\ \Delta[n] & \xrightarrow{v} & X \end{array}$$

there is a *unique* morphism $s\colon \Delta[n]\to E$ satisfying

$$p\circ s=v, \qquad s\circ i=u.$$

The objects of R/X are the *coverings* of X, and R/X is the category of coverings over X.

Let $f\colon X'\to X$ be a morphism of $\Delta^{\circ}\mathscr{E}$, and p be a covering of X. The cartesian square

$$\begin{array}{ccc} X'\times_X E & \longrightarrow & E \\ {\scriptstyle p'}\big\downarrow & & \big\downarrow{\scriptstyle p} \\ X' & \xrightarrow{f} & X \end{array}$$

defines a covering p' of X'. The correspondence $\mathsf{R}/f\colon \mathsf{R}/X\to\mathsf{R}/X'$ so defined is a covariant functor. We will sometimes write $f^{-1}(p)$ for the morphism p'. The functor R/f is the *change of base* functor.

Consider in particular a vertex $a\in X_0$, and let $\tilde{a}\colon \Delta[0]\to X$ be the singular simplex defined by a. Since there is only one morphism $\Delta[n]\to\Delta[0]$ for each n, it follows from the definitions that each simplex of a covering E of $\Delta[0]$ is determined by one of its vertices, and hence that we have $E=Sk^{\circ}E$. Thus we can identify the category $\mathsf{R}/\Delta[0]$ with the category of sets, and consider that the range of the functor $\mathsf{R}/\tilde{a}$ is $\mathscr{E}$. We will say that $\mathsf{R}/\tilde{a}$ is the *fibre functor* of R/X over a.

2.2. We can give another description of R/X, more consistent with the usual definitions of coverings in $\mathscr{T}\!op$. It can be stated as follows:

Proposition: Let $p\colon E\to X$ be a morphism of $\Delta^{\circ}\mathscr{E}$. Then p is a covering if and only if it is a locally trivial morphism with discrete fibres.

The proof is easy and is left to the reader.

2.3. Let X be a simplicial set, and G a groupoid. In this paragraph, we will study how the functors Π and D, defined in II, 7.1, act on R/X and R/G. We note first that Π induces in a natural way a functor Π_X: $\Delta^\circ\mathscr{E}/X \to \mathscr{G}r/\Pi X$ and that D induces similarly a functor D_G: $\mathscr{G}r/G \to \Delta^\circ\mathscr{E}/DG$. We still write Π_X and D_G for the restrictions of these functors to R/X and R/G.

Proposition: The functors Π_X and D_G transform coverings into coverings. Hence Π_X and D_G induce functors $\mathsf{R}/X \to \mathsf{R}/\Pi X$ and $\mathsf{R}/G \to \mathsf{R}/DG$, which will still be denoted by the same symbols.

Proof:

2.3.1. Let p: $R \to G$ be a covering, and let

$$\begin{array}{ccc} \Delta[0] & \longrightarrow & DR \\ \downarrow & & \downarrow{\scriptstyle Dp} \\ \Delta[n] & \longrightarrow & DG \end{array} \tag{1}$$

be a commutative diagram. To this diagram corresponds, by adjunction, the following diagram of $\mathscr{G}r$:

$$\begin{array}{ccc} Sc[0] & \longrightarrow & R \\ \downarrow & & \downarrow{\scriptstyle p} \\ Sc[n] & \longrightarrow & G \end{array} \tag{2}$$

This diagram can be completed "commutatively" in a unique way by a morphism $Sc[n] \to R$ (induction on n): but this means that (1) can be completed "commutatively" in a unique way by a morphism $\Delta[n] \to DR$, i.e. that Dp is a covering of DG.

2.3.2. Let p: $E \to X$ be a covering of X and let

$$\begin{array}{ccc} Sc[0] & \longrightarrow & \Pi E \\ \downarrow & & \downarrow{\scriptstyle \Pi p} \\ Sc[1] & \longrightarrow & \Pi X \end{array} \tag{3}$$

be a commutative diagram. Having (3) is the same as having an object e of ΠE whose image under Πp is one of the end points of a given morphism f of ΠX. We have to show that there is a unique morphism of ΠE such that one of its end points is e, and whose image under Πp is f.

In order to prove the existence of the lifting, we note that f can be represented by a morphism ξ: $I_n \to X$ (see II, 7.2); hence we have a commutative diagram

$$\begin{array}{ccc} \Delta[0] & \xrightarrow{\tilde{e}} & E \\ {\scriptstyle e_0}\downarrow & & \downarrow{\scriptstyle p} \\ I_n & \xrightarrow{\xi} & X \end{array}$$

This diagram can be completed "commutatively" is a unique way by a morphism $\eta\colon I_n \to E$. The morphism of ΠE associated with η is a lifting g of f.

Consider now a morphism $\xi'\colon I_p \to X$ which also represents f. As above, we associate with it a unique morphism $\eta'\colon I_p \to E$, and a morphism g' of ΠE which is a lifting of f. It remains to be shown that $g' = g$. But, considering the description of the relations in the Poincaré groupoid of a simplicial set (II, 7.1), it is sufficient to give the demonstration for the following case:

Consider a morphism $\sigma\colon \Delta[2] \to X$, and write $\alpha\colon I_2 \to \Delta[2]$ $\beta\colon I_1 = \Delta[1] \to \Delta[2]$ for the morphisms associated with the increasing maps of Fig. 107

$$\begin{array}{ccccc} 0<1>2 & & & & 0<1 \\ \downarrow\ \ \downarrow\ \ \downarrow & & \text{and} & & \downarrow\ \ \downarrow \\ 0<2>1 & & & & 0<1 \end{array}$$

Fig. 107

(see II, 5.1). Set $\xi = \sigma \circ \alpha$ and $\xi' = \sigma \circ \beta$. Then the commutative diagram

$$\begin{array}{ccc} \Delta[0] & \xrightarrow{\tilde{e}} & E \\ \downarrow & & \downarrow p \\ \Delta[2] & \xrightarrow[\sigma]{} & X \end{array}$$

can be completed "commutatively" by a morphism $\tau\colon \Delta[2] \to E$, and hence we have $\eta = \tau \circ \alpha$, $\eta' = \tau \circ \beta$, and consequently $g = g'$ (see II, 7.1).

2.4. We will see now that the functor Π_X is an equivalence $\mathsf{R}/X \simeq \mathsf{R}/\Pi X$. The adjunction morphism $\Psi\colon \mathrm{Id}\, \Delta^\circ\mathscr{E} \to D\Pi$ defines for each simplicial set X a morphism $\Psi X\colon X \to D\Pi X$ and hence a change of base functor

$$\mathsf{R}/\Psi X\colon \mathsf{R}/D\Pi X \to \mathsf{R}/X.$$

Write D_X for the composed functor $(\mathsf{R}/\Psi X) \circ D_{\Pi X}\colon \mathsf{R}/\Pi X \to \mathsf{R}/X$. We see easily that Π_X is left adjoint to D_X. Moreover, if $a \in X_0$, the diagram

$$\begin{array}{ccc} \mathsf{R}/X & \underset{\mathsf{D}_X}{\overset{\Pi_X}{\rightleftarrows}} & \mathsf{R}/\Pi X \\ & \searrow \quad \swarrow & \\ & \mathscr{E} & \end{array}$$

where the oblique arrows represent the fibre functors over a, is commutative up to isomorphism. It follows that the adjunction morphism

$$\mathrm{Id}(\mathsf{R}/X) \to \mathsf{D}_X \Pi_X$$

induces an isomorphism on the zero-skeleton for each covering of X, and that the adjunction morphism

$$\Pi_X \mathsf{D}_X \to \mathrm{Id}(\mathsf{R}/\Pi X)$$

induces an isomorphism on objects for each covering of ΠX. Hence theorem 2.4.1, which was our ain, follows from proposition 2.4.2 below (the proof of this proposition is left to the reader).

2.4.1. *Theorem: The functor* D_X *and* Π_X *define an equivalence between the category* R/X *and* $\mathsf{R}/\Pi X$.

2.4.2. *Proposition: Let* X *be a simplicial set (resp. let* G *be a groupoid) and let* f *be a morphism of* R/X *(resp.* R/G*); if* f *induces an isomorphism on* 0*-skeletons (resp. on objects),* f *is an isomorphism.*

2.5. Now let $f\colon X' \to X$ be a morphism of $\Delta^\circ\mathscr{E}$. The change of base functors induce two diagram (Fig. 108)

$$(1)\quad \begin{array}{ccc} \mathsf{R}/X & \xrightarrow{\Pi_X} & \mathsf{R}/\Pi X \\ \downarrow{\scriptstyle \mathsf{R}/f} & & \downarrow{\scriptstyle \mathsf{R}/\Pi f} \\ \mathsf{R}/X' & \xrightarrow[\Pi_{X'}]{} & \mathsf{R}/\Pi X' \end{array} \qquad (2)\quad \begin{array}{ccc} \mathsf{R}/X & \xrightarrow{\mathsf{D}_X} & \mathsf{R}/\Pi X \\ \downarrow{\scriptstyle \mathsf{R}/f} & & \downarrow{\scriptstyle \mathsf{R}/\Pi f} \\ \mathsf{R}/X' & \xrightarrow[\mathsf{D}_{X'}]{} & \mathsf{R}/\Pi X' \end{array}$$

Fig. 108

Diagram (2) is commutative up to isomorphism, because D commutes with inverse limits, and hence with fibred products. Since (Π_X, D_X) and $(\Pi_{X'}, \mathsf{D}_{X'})$ are pairs formed by equivalences quasi inverse to each other, the same holds for (1).

2.6. It follows from 2.5 that R/f is an equivalence if and only if $\mathsf{R}/\Pi f$ is an equivalence. But we saw in 1.4 that $\mathsf{R}/\Pi f$ is an equivalence if and only Πf is an equivalence; when X and X' are connected, this last statement means that f induces an isomorphism between the Poincaré groups:

Theorem: Let $f\colon X' \to X$ *be a morphism of connected simplicial sets. Then* R/f *is an equivalence if and only if* f *induces an isomorphism between the Poincaré groups of* X' *and* X.

3. Simplicial Coverings and Topological Coverings

3.1. Recall that a morphism $\boldsymbol{p}\colon \boldsymbol{E} \to \boldsymbol{X}$ of $\mathscr{T}\!op$ is said to be *locally trivial with fibre* $\boldsymbol{F}$ is, for each point $\boldsymbol{x}$ of $\boldsymbol{X}$, there is an open subset $\boldsymbol{U}$ of $\boldsymbol{X}$ containing $\boldsymbol{x}$ and a homeomorphism from $\boldsymbol{U}\times\boldsymbol{F}$ onto $\boldsymbol{p}^{-1}(\boldsymbol{U})$ such that the diagram

$$\begin{array}{ccc} \boldsymbol{U}\times\boldsymbol{F} & \xrightarrow{\simeq} & \boldsymbol{p}^{-1}(\boldsymbol{U}) \\ & {\scriptstyle \text{proj.}}\searrow \quad \swarrow{\scriptstyle \boldsymbol{p}} & \\ & \boldsymbol{U} & \end{array}$$

Fig. 109

is commutative (see III, 4.1). Thus, when $\boldsymbol{E}$, $\boldsymbol{X}$ and $\boldsymbol{F}$ are Kelley spaces, the definition given here does not coincide with that of III, 4.1. Note however that a locally trivial morphism of $\mathscr{K}e$ is a locally trivial morphism of $\mathscr{T}op$ if $\boldsymbol{F}$ is locally compact; this happens, in particular, when $\boldsymbol{F}$ is a discrete space.

Recall also that, in the category of topological spaces, we call *covering over the space* $\boldsymbol{X}$ any morphism of range $\boldsymbol{X}$ which is locally trivial, with discrete fibres. We denote by $\mathsf{R}/\boldsymbol{X}$ the full subcategory of $\mathscr{T}op/\boldsymbol{X}$ whose objects are the coverings.

For each continuous map $\boldsymbol{f}$: $\boldsymbol{X}'\to\boldsymbol{X}$, as in 2.1, we define, by means the fibred product, a change of base functor

$$\mathsf{R}/\boldsymbol{f}\colon \mathsf{R}/\boldsymbol{X}\to\mathsf{R}/\boldsymbol{X}'.$$

When $\boldsymbol{f}$ is the inclusion of a point $\boldsymbol{x}$ into $\boldsymbol{X}$, $\mathsf{R}/\boldsymbol{f}$ associates with the covering $\boldsymbol{p}\colon \boldsymbol{E}\to\boldsymbol{X}$ the discrete topological space $\boldsymbol{p}^{-1}(\boldsymbol{x})$ over $\{\boldsymbol{x}\}$. We say also that $\boldsymbol{p}\rightsquigarrow\boldsymbol{p}^{-1}(\boldsymbol{x})$ is the *fibre functor over* $\boldsymbol{x}$.

3.2. We will study now the way the functors S and $|\,?|$ defined in III, 1.4 transform coverings. As in paragraph 2, we see that S and $|\,?|$ induce functors

$$S_{\boldsymbol{X}}\colon \mathsf{R}/\boldsymbol{X}\to\Delta^\circ\mathscr{E}/S\boldsymbol{X} \quad\text{and}\quad |\,?|_X\colon \mathsf{R}/X\to(\mathscr{T}op)/|X|$$

and as above, there functors actually take their values in the categories $\mathsf{R}/S\boldsymbol{X}$ and R/X. Indeed:

Theorem: The image of a topological covering under the singular complex functor is a covering of $\Delta^\circ\mathscr{E}$. The geometric realization of a covering of $\Delta^\circ\mathscr{E}$ is a topological covering.

Proof: In order to prove the first part, we note that since the functors S and $|\,?|$ are adjoint, a square of $\Delta^\circ\mathscr{E}$ of type (2) is equivalent to a square of $\mathscr{T}op$ of type (1) (Fig. 110)

$$(1)\quad \begin{array}{ccc} \Delta^\circ & \xrightarrow{\boldsymbol{f}} & \boldsymbol{E} \\ {\scriptstyle \Delta^u}\downarrow & & \downarrow{\scriptstyle \boldsymbol{p}} \\ \Delta^n & \xrightarrow{\boldsymbol{y}} & \boldsymbol{X} \end{array} \qquad\qquad (2)\quad \begin{array}{ccc} \Delta[0] & \longrightarrow & S\boldsymbol{E} \\ {\scriptstyle \Delta(u)}\downarrow & & \downarrow{\scriptstyle S\boldsymbol{p}} \\ \Delta[n] & \longrightarrow & S\boldsymbol{X} \end{array}$$

Fig. 110

Suppose then that $\boldsymbol{p}$ is a covering. In order to prove that $S\boldsymbol{p}$ is a covering, it is sufficient to prove that the commutativity of (1) implies the existence and the unicity of a continuous map $\boldsymbol{s}\colon \Delta^n\to\boldsymbol{E}$ such that $\boldsymbol{p}\circ\boldsymbol{s}=\boldsymbol{y}$ and $\boldsymbol{f}=\boldsymbol{s}\circ\Delta^u$. To that end we can restrict ourselves first to the case $n=1$ (path lifting). When n is arbitrary, we can then define $\boldsymbol{s}$ by lifting into $\boldsymbol{E}$ the segments of Δ^n whose origin is the image of Δ^u (notice that $\boldsymbol{p}$ is a fibration in the sense of SERRE (VII, 1.4)).

Now let $p: E \to X$ be a covering over X. By 2.2, p is a locally trivial morphism with discrete fibre over each connected component of X. Since the geometric realization of a simplicial set reduced to its 0-skeleton is a space with the discrete topology, theorem 4.2 of chapter III shows that $|p|: |E| \to |X|$ is a morphism of $\mathcal{K}e$ which is locally trivial with discrete fibre over each connected component of X.

By 3.1, such a morphism is a covering of $\mathcal{T}op$. Q.E.D.

Remark: Let $\boldsymbol{x} \in \boldsymbol{X}$ and $a \in X_0$. It follows from the proof of the theorem that the diagram of Fig. 111

$$\begin{array}{ccc} \mathsf{R}/\boldsymbol{X} & \xrightarrow[S_{\boldsymbol{X}}]{} & \mathsf{R}/S\boldsymbol{X} \\ & \searrow \quad \swarrow & \\ & \mathcal{E} & \end{array} \qquad \begin{array}{ccc} \mathsf{R}/X & \xrightarrow[|?|_X]{} & \mathsf{R}/|X| \\ & \searrow \quad \swarrow & \\ & \mathcal{E} & \end{array}$$

Fig. 111

where oblique arrows are fibre functors over $\boldsymbol{x}$ or a, are commutative up to isomorphism.

3.2.1. Let Ψ: Id $\Delta°\mathcal{E} \to S\,|\,?\,|$ be the adjunction morphism given by proposition II, 1.3. For each complex X, the morphism $\Psi X: X \to S|X|$ defines a change of base functor $\mathsf{R}/\Psi X$. Write S_X for the composition $(\mathsf{R}/\Psi X) \circ S_{|X|}$: $\mathsf{R}/|X| \to \mathsf{R}/X$. This functor is right adjoint to $|\,?\,|_X$, and moreover, we have the following theorem:

Theorem: The functors $|\,?\,|_X$ *and* S_X *define an equivalence between the categories* R/X *and* $\mathsf{R}/|X|$.

Proof: The proof is similar to that of theorem 2.4. It is a consequence of proposition 2.4 and of a similar proposition about topological coverings over a locally connected space [recall that the geometric realization of a complex is locally connected (III, 1.10)]:

3.2.2. *Proposition: Let*

$$\begin{array}{ccc} \boldsymbol{E} & \xrightarrow{\boldsymbol{f}} & \boldsymbol{E}' \\ & {\scriptstyle \boldsymbol{p}}\searrow \quad \swarrow{\scriptstyle \boldsymbol{p}'} & \\ & \boldsymbol{X} & \end{array}$$

be a morphism of $\mathsf{R}/\boldsymbol{X}$, *where* $\boldsymbol{X}$ *is a locally connected topological space. If the restriction of* $\boldsymbol{f}$ *to fibres is an isomorphism,* $\boldsymbol{f}$ *is an isomorphism.*

3.2.3. *Remark:* Let f: $X' \to X$ be a morphism of $\Delta°\mathcal{E}$. Since the functors $|\,?\,|$ and S commute with fibred products (III), the square of Fig. 112

$$\begin{array}{ccc} \mathsf{R}/X & \xrightarrow{|?|_X} & \mathsf{R}/|X| \\ {\scriptstyle \mathsf{R}/f}\downarrow & & \downarrow{\scriptstyle \mathsf{R}/|f|} \\ \mathsf{R}/X' & \xrightarrow{|?|_{X'}} & \mathsf{R}/|X'| \end{array} \qquad \begin{array}{ccc} \mathsf{R}/X & \xleftarrow{\mathsf{S}_X} & \mathsf{R}/|X| \\ {\scriptstyle \mathsf{R}/f}\downarrow & & \downarrow{\scriptstyle \mathsf{R}/|f|} \\ \mathsf{R}/X' & \xleftarrow{\mathsf{S}_{X'}} & \mathsf{R}/|X'| \end{array}$$

Fig. 112

are commutative up to functor isomorphism.

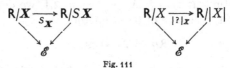

3.3. We saw in paragraph 3.2.1 that $|\,?\,|_X\colon \mathsf{R}/X \to \mathsf{R}/|X|$ is an equivalence. Now we intend to show that, if the space $\boldsymbol{X}$ satisfies certain conditions, then $S_{\boldsymbol{X}}\colon \mathsf{R}/\boldsymbol{X} \to \mathsf{R}/S\boldsymbol{X}$ is also an equivalence. In order to do this, *we will show first that the composition* $\Pi \circ S\colon \mathcal{T}op \to \mathcal{G}r$ *associates with each space* $\boldsymbol{X}$ *the Poincaré groupoid* $G(\boldsymbol{X})$, which *is traditionally defined as follows:*

The diagram scheme subordinated to $G(\boldsymbol{X})$ is a quotient of the diagram scheme $d_1, d_0\colon (S\boldsymbol{X})_1 \rightrightarrows (S\boldsymbol{X})_0$ where d_1 (resp. d_0) denotes the domain map (resp. the range map). The objects of $G(\boldsymbol{X})$ coincide with the elements of $(S\boldsymbol{X})_0$, i.e. with the points of $\boldsymbol{X}$; the set of morphisms of $G(\boldsymbol{X})$ is the quotient of $(S\boldsymbol{X})_1$ by an equivalence relation such that $\boldsymbol{f} \sim \boldsymbol{g}$ if and only if conditions 1) and 2) are satisfied: 1) $\boldsymbol{f}(0) = \boldsymbol{g}(0)$ and $\boldsymbol{f}(1) = \boldsymbol{g}(1)$; 2) there is a map $\boldsymbol{h}\colon \Delta^1 \times \Delta^1 \to \boldsymbol{X}$ such that $\boldsymbol{h}(0, ?) = \boldsymbol{f}$, $\boldsymbol{h}(1, ?) = \boldsymbol{g}$, $\boldsymbol{h}(t, e) = \boldsymbol{f}(e) = \boldsymbol{g}(e)$ for all $t \in T$ and $e = 0, 1$[1].

If $\bar{\boldsymbol{f}}$ and $\bar{\boldsymbol{g}}$ are two composable morphisms of $G(\boldsymbol{X})$, and if $\boldsymbol{f}$ and $\boldsymbol{g}$ are representatives of $\bar{\boldsymbol{f}}$ and $\bar{\boldsymbol{g}}$ in $(S\boldsymbol{X})_1$, $\bar{\boldsymbol{g}} \circ \bar{\boldsymbol{f}}$ is by definition the equivalence class of the map $\boldsymbol{g} * \boldsymbol{f}$ defined by the following equations: $\boldsymbol{g} * \boldsymbol{f}(t) = \boldsymbol{f}(2t)$ if $0 \leqq t \leqq \frac{1}{2}$ and $\boldsymbol{g} * \boldsymbol{f}(t) = g(2t-1)$ if $\frac{1}{2} \leqq t \leqq 1$. With this definition of composition, $G(\boldsymbol{X})$ is a category, and even a groupoid, because each morphism $\bar{\boldsymbol{f}}$, with representative $\boldsymbol{f}$, has an inverse which is the equivalence class of the map $\boldsymbol{f}^-$ defined by $\boldsymbol{f}^-(t) = \boldsymbol{f}(1-t)$.

We can prove then that $G(\boldsymbol{X})$ is isomorphic to $\Pi S\boldsymbol{X}$, by an argument similar to that of IV, 4.2.

We call *Poincaré group* of $\boldsymbol{X}$ at $\boldsymbol{x}$, and we denote by $\Pi_1(\boldsymbol{X}, \boldsymbol{x})$, the group $\Pi_1(S\boldsymbol{X}, \boldsymbol{x})$. It follows from the preceding remarks that this definition coincide with the classical definition, and also with that of VII, 2.5.

3.4. Let $(\Pi S\boldsymbol{X})^\circ\mathcal{E}$ be the category of local systems on $\Pi S\boldsymbol{X}$: by 3.3 it is the category of local systems on $\boldsymbol{X}$, as usually defined in topology. Write $\mathsf{D}_{\boldsymbol{X}}$ for the composition

$$\mathsf{R}/\boldsymbol{X} \xrightarrow{S_{\boldsymbol{X}}} \mathsf{R}/S\boldsymbol{X} \xrightarrow{\Pi_{S\boldsymbol{X}}} \mathsf{R}/\Pi S\boldsymbol{X} \xrightarrow{\ \mathsf{L}\ } (\Pi S\boldsymbol{X})^\circ\mathcal{E}.$$

Since $\Pi_{S\boldsymbol{X}}$ and L are equivalences of categories, the study of $S_{\boldsymbol{X}}$ reduces to that of $\mathsf{D}_{\boldsymbol{X}}$.

The functor $\mathsf{D}_{\boldsymbol{X}}$ can be given explicitely as follows:

Let $\boldsymbol{p}\colon \boldsymbol{E} \to \boldsymbol{X}$ be a covering of $\boldsymbol{X}$ and u a 1-simplex of $\boldsymbol{X}$ with starting point $\boldsymbol{x}$ and end point $\boldsymbol{y}$. For each point $\boldsymbol{\omega}$ of the fibre $\boldsymbol{p}^{-1}(\boldsymbol{y})$, there is a unique 1-simplex v of $\boldsymbol{E}$ such that we have $\boldsymbol{p} \circ v = u$, $v(1) = \boldsymbol{\omega}$. Let $v(0) = u^*(\boldsymbol{\omega})$; we define thus a map $u^*\colon \boldsymbol{p}^{-1}(\boldsymbol{y}) \to \boldsymbol{p}^{-1}(\boldsymbol{x})$. If u' is another 1-simplex of $\boldsymbol{X}$ such that u and u' represent the same morphism of $\Pi S\boldsymbol{X}$, the maps u^* and u'^* are equal.

[1] We identify here the interval $\boldsymbol{I} = [0, 1]$ of $\boldsymbol{R}$ with Δ^1 be means of the map $t \leadsto (0, t, 1)$.

The functor $\mathsf{D}_{\boldsymbol{X}}$ is then given by the equalities:

$$(\mathsf{D}_{\boldsymbol{X}}\boldsymbol{p})(\boldsymbol{x}) = \boldsymbol{p}^{-1}(\boldsymbol{x}), \qquad (\mathsf{D}_{\boldsymbol{X}}\boldsymbol{p})(u) = u^{*}.$$

3.5. *Proposition: If the space* $\boldsymbol{X}$ *is locally pathwise connected, the functor* $S_{\boldsymbol{X}}$ *is fully faithful.*

3.6. *Theorem: If* $\boldsymbol{X}$ *is a locally pathwise connected space, such that each point has a simply connected open neighborhood, the functor* $S_{\boldsymbol{X}}$: $\mathsf{R}/\boldsymbol{X} \to \mathsf{R}/S\boldsymbol{X}$ *is an equivalence of categories.*

In both cases, it is sufficient to prove the corresponding property for $\mathsf{D}_{\boldsymbol{X}}$: in view of the description of this functor given in 3.4, the demonstrations are then "well known", and are left to the reader.

3.7. Let us return now to the situation of theorem 3.2.1: S_X is an equivalence of categories. But S_X was defined by the equality

$$\mathsf{S}_X = (\mathsf{R}/\Psi X) \circ S_{|X|}$$

and we proved in chapter III that $|X|$ is locally pathwise connected, and that each point of $|X|$ has a contractible neighborhood. Hence the topological space $|X|$ satisfies the hypothesis of theorem 3.6. Hence the functor $S_{|X|}$, and also $\mathsf{R}/\Psi X$, are equivalences of categories.

Let us apply the result of 2.6; we obtain then the following particular case of MILNOR's theorem:

3.7.1. *Theorem: Let* X *be a simplicial set,* x_0 *a vertex of* X*, and* Ψ: $\mathrm{Id}\,\Delta^{\circ}\mathscr{E} \to S|\,?\,|$ *an adjunction morphism. The morphism* ΨX *induces an isomorphism of Poincaré groups*

$$\Pi_1(\Psi X, x_0) : \Pi_1(X, x_0) \simeq \Pi_1(S|X|, x_0).$$

3.7.2. *Corollary:* ("Van Kampen" for geometric realizations.) *Let* X *be a connected simplicial set, and* A *and* B *connected simplicial subsets of* X *such that* $A \cup B = X$ *and* $A \cap B$ *is connected. Then if* $x_0 \in (A \cap B)_0$*, we have a canonical isomorphism*

$$\Pi_1(|X|, x_0) \approx \Pi_1(|A|, x_0) \overset{\Pi_1(|A\cap B|,\, x_0)}{\amalg} \Pi_1(|B|, x_0).$$

This is an immediate consequence of 3.7.1 and II, 7.4.

Appendix Two

The Homology Groups of a Simplicial Set

1. A Theorem of Eilenberg

1.1. Let X be a simplicial set and $C_n X$ the free abelian group generated by the n-simplices of X. The face operators d_i^n: $X_n \to X_{n-1}$ induce a homomorphism of abelian groups δ_i^n: $C_n X \to C_{n-1} X$. Setting

$\delta_n = \sum_{i=0}^{i=n} (-1)^i \delta_i^n$, we obtain a differential abelian group

$$\ldots C_2X \to C_1X \to C_0X,$$

which will be denoted by C_*X. By definition, the n-th homology group of C_*X will be written H_nX and called *the* n-th *homology group* of X. This group depends functorially on X.

If Y is a topological space, it is a well-known fact that the n-th homology group of the singular complex SY is, by definition, the n-th singular homology group of Y; it is denoted by H_nY and depends functorially on Y; we will see that the theorem of MILNOR (VII, 1.3) implies that, for every simplicial set X, H_nX is canonically isomorphic to $H_n|X|$. Chapter VII gives therefore as a premium a new proof for the well-known theorem of EILENBERG.

1.2. Let $\Psi X\colon X \to S|X|$ be the morphism which is canonically associated with the adjunction isomorphism

$$\mathcal{T}op(|Z|, Y) \to \Delta^\circ\mathcal{E}(Z, SY)$$

of III, 1.4. By MILNOR's theorem (VII, 3.1), ΨX induces an isomorphism of the homotopic category $\mathcal{H}$. In order to show that H_nX is identified with $H_n|X|$, it is sufficient to prove that $H_n(\Psi X)$ is invertible, or that the functor H_n may be factorised through $\mathcal{H}$.

By Lemma 1.4 below, we know that $H_nf = H_ng$ if f and g are connected by a homotopy. This means that H_n may be factorised through the category $\overline{\Delta^\circ\mathcal{E}}$ of complexes modulo homotopy. As $\mathcal{H}$ is a category of fractions of $\overline{\Delta^\circ\mathcal{E}}$, it remains to prove that H_ns is invertible when s is an anodyne extension (IV, 2.1.4).

Let us write Σ for the set of monomorphisms of $\Delta^\circ\mathcal{E}$ such that H_ns is invertible. It is clear that Σ contains the inclusion of $\Lambda^k[n]$ into $\Delta[n]$ (IV, 2), because that inclusion induces an isomorphism in $\overline{\Delta^\circ\mathcal{E}}$. Moreover, Σ is saturated in the sense of IV, 2.1: in fact, conditions (i), (iii) and (iv) are trivial, so we only have to prove (ii). With the notations of IV, 2.1, the square

$$\begin{array}{ccc} C_*X & \longrightarrow & C_*Y \\ {\scriptstyle C_*\xi}\downarrow & & \downarrow{\scriptstyle C_*\eta} \\ C_*X' & \longrightarrow & C_*Y' \end{array}$$

is cocartesian, because the functor $X \rightsquigarrow C_*X$ commutes with direct limits. Our assertion follows therefore from Lemma 1.3 below.

1.3. *Lemma: Let*

$$\begin{array}{ccc} A & \xrightarrow{\gamma} & B \\ {\scriptstyle \alpha}\downarrow & & \downarrow{\scriptstyle \beta} \\ C & \xrightarrow[\delta]{} & D \end{array}$$

be a cocartesian square of differential abelian groups, such that α is a monomorphism. If $H_n\alpha$ is invertible for every n, the same is true for $H_n\beta$.

In fact, the exact sequence

$$0 \longrightarrow A \xrightarrow{(\gamma,\,-\alpha)} B \oplus C \xrightarrow{(\beta,\,\delta)} D \longrightarrow 0$$

induces an infinite exact sequence

$$\ldots H_n A \to H_n B \oplus H_n C \to H_n D \to H_{n-1} A \ldots .$$

As $H_n\alpha$ and $H_{n-1}\alpha$ are invertible, we have an exact sequence

$$0 \longrightarrow H_n A \xrightarrow{(H_n\gamma,\,-H_n\alpha)} H_n B \oplus H_n C \xrightarrow{(H_n\beta,\,H_n\delta)} H_n D \longrightarrow 0.$$

This again means that the square

$$\begin{array}{ccc} H_n A & \xrightarrow{H_n\gamma} & H_n B \\ {\scriptstyle H_n\alpha}\downarrow & & \downarrow{\scriptstyle H_n\beta} \\ H_n C & \xrightarrow[H_n\delta]{} & H_n D \end{array}$$

is cocartesian, so that $H_n\beta$ is invertible if $H_n\alpha$ is.

1.4. *Lemma: Homotopic morphisms of simplicial sets $f, g\colon X \rightrightarrows Y$ induce homotopic maps of differential groups $C_* f, C_* g\colon C_* X \rightrightarrows C_* Y$.*

In fact, it is clear that we can reduce the proof to the case where f, g are connected by a homotopy h; in that case, $f = h\varepsilon_0$ and $g = h\varepsilon_1$, where ε_0 and ε_1 stand for $\Delta(\partial_1^1) \times X$ and $\Delta(\partial_1^0) \times X$ (IV, 1.1). Thus, it is sufficient to prove that $C_*(\varepsilon_0)$ and $C_*(\varepsilon_1)$ are homotopic, i.e. to find group homomorphisms $s_n\colon C_n X \to C_{n+1}(\Delta[1] \times X)$ such that $\delta s_n + s_{n-1}\delta = C_n(\varepsilon_1) - C_n(\varepsilon_0)$. Let $\tau_i\colon [n+1] \to [1]$ be the non-decreasing map which takes $i+1$ times the value 0, so that $\tau_i \in C_{n+1}(\Delta[1])$. Notice then that $C_* X$ is a simplicial group, and just set

$$s_n x = \sum_{i=0}^{i=n} (-1)^i (\tau_i, x s_i^n x),$$

with notations from II, 2.4.

The definition of s_n is perhaps more understandable if we notice that s_n is a morphim from the functor $C_n\colon \Delta^\circ\mathscr{E} \to \mathscr{Ab}$ to the functor $X \rightsquigarrow C_{n+1}(\Delta[1] \times X)$. As $C_n X$ is the free abelian group $X_n^{(\mathbb{Z})}$ generated by X_n, s_n may be identified with a morphism of set functors from $?_n$ to C_{n+1}. But, as $?_n$ is isomorphic to $\Delta^\circ\mathscr{E}(\Delta[n], ?)$, s_n is canonically associated with an element $s_n' \in C_{n+1}(\Delta[1] \times \Delta[n]) \simeq (\mathscr{Or}([n+1], [1] \times [n]))^{(\mathbb{Z})}$, with notations from II, 5. This s'_n is the alternating sum of the increasing maps $\varrho_0, \varrho_1, \varrho_2, \ldots$ represented by the following chains (II, 5):

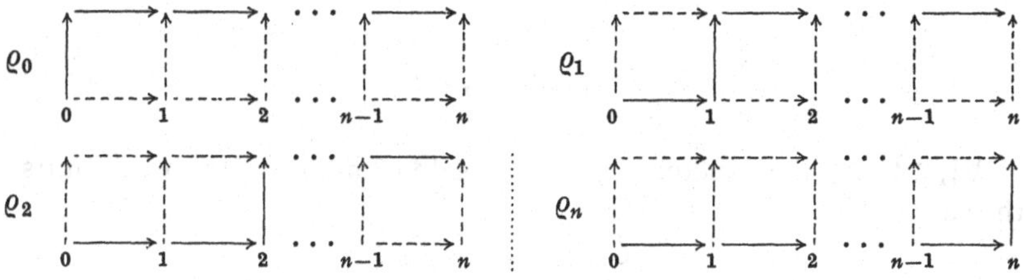

Fig. 113

2. The Reduced Homology Groups of a Pointed Simplicial Set

2.1. Let (E, e) be a pointed set. The free abelian group $(E, e)^{(\mathbb{Z})}$ generated by (E, e) is, by definition, the quotient of the free abelian group $E^{(\mathbb{Z})}$, generated by E, by the relation $e=0$.

If (X, x) is a pointed simplicial set, we will write $\dot{C}_* X$ for the differential group whose n-th component $\dot{C}_n X$ is the free abelian group generated by the pointed set (X_n, x); the boundary operator is defined "as in 1.1". If $n \geqq 0$, the n-th homology group of $\dot{C}_n X$ will be denoted by $\dot{H}_n X$ and called the n-th *reduced homology group of* (X, x).

The groups $\dot{H}_n X$ and $H_n X$ are simply related: let Z_* be the differential group such that $Z_n = 0$ if $n < 0$, $Z_n = \mathbb{Z}$ if $n \geqq 0$, the boundary operator $d_n \colon Z_n \to Z_{n-1}$ being defined as follows: $d_n = 0$ if $n \leqq 0$ or if n is odd; $d_n = \operatorname{Id} \mathbb{Z}$ if $n > 0$ is even. We then have by definition a canonical exact sequence

$$0 \to Z_* \to C_* X \to \dot{C}_* X \to 0,$$

inducing the usual infinite sequence of homology groups. As $H_n Z_* = 0$ if $n \neq 0$ and $H_0 Z = \mathbb{Z}$, we see that

$$H_n X \simeq \dot{H}_n X \quad \text{if} \quad n > 0$$

and that there is an exact sequence

$$0 \to \mathbb{Z} \to H_0 X \to \dot{H}_0 X \to 0,$$

where $\mathbb{Z}$ is identified with the subgroup of $H_0 X$ generated by the connected component of X containing x (notice that $H_0 X$ is identified with the free abelian group generated by the connected components of X).

2.2. Consider now a morphism $f: X \to Y$ of $.\Delta^\circ\mathscr{E}$ and the cocartesian square

$$\begin{array}{ccc} X & \xrightarrow{f} & Y \\ {\scriptstyle\varepsilon}\downarrow & & \downarrow{\scriptstyle if} \\ \Delta[1] \wedge X & \xrightarrow[\eta]{} & Cf \end{array}$$

of VI, 2.1. As the functors $\dot{C}_n: .\Delta^\circ\mathscr{E} \to \mathscr{Ab}$ commute with direct limits, the square

$$\begin{array}{ccc} \dot{C}_* X & \longrightarrow & \dot{C}_* Y \\ \downarrow & & \downarrow \\ \dot{C}_*(\Delta[1] \wedge X) & \longrightarrow & \dot{C}_*(Cf) \end{array}$$

is also cocartesian and induces by 1.3 an infinite exact sequence

$$\ldots \dot{H}_{n+1}(Cf) \xrightarrow{\partial_{n+1} f} \dot{H}_n X \xrightarrow{(\dot{H}_n f,\, -\dot{H}_n \varepsilon)} \dot{H}_n Y \oplus \dot{H}_n(\Delta[1] \wedge X) \xrightarrow{(\dot{H}_n if,\, \dot{H}_n \eta)} \dot{H}_n Cf \ldots, n \geqq$$

As $\Delta[1] \wedge X$ is isomorphic to $\Delta[0]$ in $\overline{\Delta^\circ\mathscr{E}}$, we have $\dot{H}_n(\Delta[1] \wedge X) = 0$ and the infinite exact sequence

$$(f_*) \ldots \dot{H}_{n+1}(Cf) \xrightarrow{\partial_{n+1} f} \dot{H}_n X \xrightarrow{\dot{H}_n f} \dot{H}_n Y \xrightarrow{\dot{H}_n if} \dot{H}_n(Cf) \ldots, \qquad n \geqq 0.$$

Moreover, the commutative square

$$\begin{array}{ccc} X & \xrightarrow{f} & Y \\ {\scriptstyle \mathrm{Id}\, X}\downarrow & & \downarrow \\ X & \xrightarrow{0} & \Delta[0] \end{array}$$

induces a morphism between the corresponding infinite sequences

$$\begin{array}{lcccccc} (f_*) \ldots \dot{H}_{n+1}(Cf) & \xrightarrow{\partial_{n+1} f} & \dot{H}_n X & \longrightarrow & \dot{H}_n Y & \longrightarrow & \dot{H}_n(Cf) \\ \qquad \downarrow{\scriptstyle \dot{H}_{n+1} jf} & & \downarrow{\wr} & & \downarrow & & \downarrow{\scriptstyle \dot{H}_n(jf)} \\ (0_*) \ldots \dot{H}_{n+1}(\Sigma X) & \xrightarrow[\partial_{n+1} 0]{} & \dot{H}_n X & \longrightarrow & \dot{H}_n \Delta[0] & \longrightarrow & \dot{H}_n(\Sigma X) \end{array}$$

In particular, we have isomorphisms $\dot{H}_{n+1}(\Sigma X) \underset{\partial_{n+1} 0}{\xrightarrow{\sim}} \dot{H}_n(X)$, $n \geqq 0$ and commutative triangles

$$\begin{array}{ccc} \dot{H}_{n+1}(Cf) & \searrow^{\partial_{n+1} f} & \\ \downarrow{\scriptstyle \dot{H}_{n+1}(jf)} & & \dot{H}_n(X) \\ \dot{H}_{n+1}(\Sigma X) & \nearrow_{\partial_{n+1} 0}^{\sim} & \end{array}$$

3. The Spectral Sequence of Direct Limits

In the sequel, $\mathcal{M}$ stands for an abelian category with exact infinite direct sums. We intend to give a brief account of the spectral sequence of a fibration. This paragraph is preparatory.

3.1. Let $\mathcal{A}$ be a small category. Recall that we have associated with $\mathcal{A}$ a simplicial set $D\mathcal{A}$ (II, 4.1), which may be described as follows: $(D\mathcal{A})_0 = \mathfrak{Ob}\,\mathcal{A}$ and, if $n > 0$, $(D\mathcal{A})_n$ is the set of all n-sequences of $\mathcal{A}$, i.e. the set of the following diagrams of $\mathcal{A}$

$$a_n \xleftarrow{\alpha_n} a_{n-1} \ldots a_2 \xleftarrow{\alpha_2} a_1 \xleftarrow{\alpha_1} a_0 .$$

Such a diagram will be denoted by $\alpha = (\alpha_n, \ldots, \alpha_2, \alpha_1)$; the face and degeneracy operators are defined by the formulas

$$d_0^1 \alpha = \mathfrak{r}\alpha, \qquad d_1^1 \alpha = \mathfrak{d}\alpha$$

$$d_0^n(\alpha_n, \ldots, \alpha_2, \alpha_1) = (\alpha_n, \ldots, \alpha_2)$$

$$d_i^n(\alpha_n, \ldots, \alpha_1) = (\alpha_n, \ldots, \alpha_{i+2}, \alpha_{i+1}\alpha_i, \alpha_{i-1}, \ldots, \alpha_1), \qquad 0 < i < n$$

$$d_n^n(\alpha_n, \alpha_{n-1}, \ldots, \alpha_1) = (\alpha_{n-1}, \ldots, \alpha_1)$$

and

$$s_i^n(\alpha_n, \ldots, \alpha_1) = (\alpha_n, \ldots, \alpha_{i+1}, \mathrm{Id}\, a_i, \alpha_i, \ldots, \alpha_1)$$

with the preceding notations.

3.2. Consider now the category $\mathcal{A}\mathcal{M}$ of functors from $\mathcal{A}$ to $\mathcal{M}$. For every object L of $\mathcal{A}\mathcal{M}$ and each n-sequence α, we set $L\alpha = L a_0$, the notations being those of 3.1. We define then a simplicial object $C_*(\mathcal{A}, L)$ of $\mathcal{M}$ by setting $C_n(\mathcal{A}, L) = \coprod_\alpha L\alpha$, where α runs through the n-sequences of $\mathcal{A}$; if we identify $L\alpha$ with a subobject of $C_n(\mathcal{A}, L)$, the restrictions of the operators d_0^n, d_i^n $(i > 0)$, s_i^n to $L\alpha$ are respectively the following composed morphisms

$$L\alpha \xrightarrow{L\alpha_1} L(d_0^n \alpha) \xrightarrow{in_{d_0\alpha}} C_{n-1}(\mathcal{A}, L)$$

$$L\alpha \xrightarrow{\mathrm{Id}} L(d_i^n \alpha) \xrightarrow{in_{d_i\alpha}} C_{n-1}(\mathcal{A}, L)$$

$$L\alpha \xrightarrow{\mathrm{Id}} L(s_i^n \alpha) \xrightarrow{in_{s_i\alpha}} C_{n+1}(\mathcal{A}, L).$$

If we set $\delta_n = \sum_{i=0}^{i=n} (-1)^i d_i^n$, $(C_*(\mathcal{A}, L), \delta_*)$ becomes a differential group, whose homology objects will be denoted by $H_n(\mathcal{A}, L)$.

3.3. *Proposition: Let $L\colon \mathcal{A} \to \mathcal{M}$ be a functor from a small category $\mathcal{A}$ to an abelian category $\mathcal{M}$ with exact infinite direct sums. The notations being those of* 3.2, *$H_0(\mathcal{A}, L)$ is identified with $\varinjlim L$ and $H_n(\mathcal{A}, L)$ is identified with the n-th left satellite $\varinjlim_n^{\mathcal{A}} L$ of the functor $\varinjlim^{\mathcal{A}}\colon \mathcal{A}\mathcal{M} \to \mathcal{M}$ given by $L \rightsquigarrow \varinjlim L$.*

Indeed, $H_0(\mathscr{A}, L)$ is just the cokernel of the pair

$$d_0, d_1\colon \coprod_{\alpha\in\mathfrak{Ar}A} L\alpha \rightrightarrows \coprod_{a\in\mathfrak{Ob}A} La.$$

But this gives the usual construction of the direct limit of L. Notice then that $C_*(\mathscr{A}, L)$ is an exact functor in L, so that an exact sequence of L's gives rise to the usual infinite sequence of homology objects. This means that the $H_n(\mathscr{A}, L)$'s are part of an exact connected sequence of functors. In order to prove that this sequence is universal, it is sufficient to prove that, for each L, there is an epimorphism $p\colon M\to L$ such that $H_n(\mathscr{A}, M)=0$ if $n>0$. This is done in 3.4 below.

3.4. Let $N=(Na)_{a\in\mathfrak{Ob}\mathscr{A}}$ be a family of objects of $\mathscr{M}$, indexed by $\mathfrak{Ob}\,\mathscr{A}$. Notice first that objects $C_n(\mathscr{A}, N)$ and operators s_i^n and d_j^n ($j>0$; we do not define d_0^n here) may be defined with the help of the same formulas as in 3.2. Write then i^*N for the following functor from $\mathscr{A}$ to $\mathscr{M}$: $(i^*N)(a)=\coprod_{\mathfrak{r}\alpha=a} N(\mathfrak{d}\alpha)$, where $\alpha\colon x\to a$ runs through the morphisms of $\mathscr{A}$ with range a; if $\xi\colon a\to b$ belongs to $\mathfrak{Ar}\,\mathscr{A}$, $(i^*N)(\xi)$ induces on the summand $N(\mathfrak{d}\alpha)$ of $(i^*N)(a)$ the canonical monomorphism from $N(\mathfrak{d}\alpha)$, which is identified with the summand $N(\mathfrak{d}\xi\alpha)$ of index $\xi\alpha$ of $(i^*N)(b)$, into $(i^*N)(b)$.

We have then

$$C_n(\mathscr{A}, i^*N)=\coprod_{\alpha}(i^*N)(a_0)=\coprod_{\alpha}\coprod_{\beta}N(\mathfrak{d}\beta),$$

where α runs through the n-sequences $a_n \xleftarrow{\alpha_n} \cdots \xleftarrow{\alpha_1} a_0$ and β through the morphisms with range a_0. As (α,β) may be identified with the $(n+1)$-sequence $a_n \xleftarrow{\alpha_n} \cdots \xleftarrow{\alpha_1} a_0 \xleftarrow{\beta} \mathfrak{d}\beta$, we have the equality $C_n(\mathscr{A}, i^*N) \simeq C_{n+1}(\mathscr{A}, N)$. Moreover, an easy verification shows that the operators $'d_i^n$ and $'s_i^n$ of $C_*(\mathscr{A}, i^*N)$ are related to the operators d_i^n and s_i^n of $C_*(\mathscr{A}, N)$ by the formulas

$$'d_i^n = d_{i+1}^{n+1} \qquad 's_i^n = s_{i+1}^{n+1}.$$

Setting $'\delta_n=\sum_{i=0}^{i=n}(-1)^i\,'d_i^n$, we have

$$s_0^n\,'\delta_n + '\delta_{n+1}\,s_0^{n+1} = \mathrm{Id} \quad \text{if} \quad n>0$$

and

$$s_0^0\,d_1^1 + '\delta_1\,s_0^1 = \mathrm{Id}, \qquad d_1^1\,s_0^0 = \mathrm{Id}.$$

Thus, the existence of a homotopy operator $(s_0^1, s_0^2, \ldots)$ implies that $H_n(\mathscr{A}, i^*N)=0$ if $n>0$ and $H_0(\mathscr{A}, i^*N)=\coprod_{a\in\mathfrak{Ob}A} Na$.

Now, if L is any object of $\mathscr{A}\mathscr{M}$, let N be the family of objects La of $\mathscr{M}$ ($a\in\mathfrak{Ob}\,\mathscr{A}$), and set $M=i^*N$. For each $a\in\mathfrak{Ob}\,\mathscr{A}$, let $pa\colon (\coprod_{\mathfrak{r}\alpha=a} L\mathfrak{d}\alpha)\to La$

be the morphism whose component of index α is $L\alpha$. The morphisms $p\,a$ induce obviously a functor epimorphism $p: M \to L$; moreover, we have already seen that $H_n(\mathcal{A}, M) = 0$ if $n > 0$.

3.5. We consider now two small categories $\mathcal{A}, \mathcal{B}$ and a functor $f: \mathcal{B} \to \mathcal{A}$. We write $f_*: \mathcal{A}\mathcal{M} \to \mathcal{B}\mathcal{M}$ for the functor $L \rightsquigarrow L \circ f$. It is a well known fact that f_* admits a left adjoint functor f^*, which may be described as follows (KAN [4]): let a be any object of $\mathcal{A}$, and f/a the "left fibre of f over a", i.e. the category whose objects are the pairs (b, ξ) formed by an object b of $\mathcal{B}$ and a morphism $\xi: fb \to a$ of $\mathcal{A}$; a morphism $(b, \xi) \to (b', \xi')$ between two objects of f/a is just a morphism $\beta: b \to b'$ of $\mathcal{B}$, such that the triangle

$$\begin{array}{ccc} fb & \searrow^{\xi} & \\ {\scriptstyle f\beta}\downarrow & & a \\ fb' & \nearrow_{\xi'} & \end{array}$$

is commutative; the composition of morphisms is defined in the obvious way.

If $N: \mathcal{B} \to \mathcal{M}$ is a functor, and a an object of $\mathcal{A}$, we define $(f^* N)(a)$ as the direct limit of the functor $(b, \xi) \rightsquigarrow N b$. We may write this as follows

$$(f^* N)(a) = \varinjlim_{x \in f/a} N(pr_1 x).$$

If $\alpha: a \to a'$ is a morphism of $\mathcal{A}$, we define

$$(f^* N)(\alpha): \varinjlim_{x \in f/a} N(pr_1 x) \to \varinjlim_{y \in f/a'} N(pr_1 y)$$

as the morphism induced by the functor f/α: $(b, \xi) \rightsquigarrow (b, \alpha\xi)$ from f/a to f/a'.

For example, if $\mathcal{B}$ is the subcategory $\mathcal{A}_0$ of $\mathcal{A}$, which has the same objects as $\mathcal{A}$ and whose only morphisms are the identities, then N is simply a family of objects Na of M $(a \in \mathfrak{Ob}\,\mathcal{A})$. If f is the inclusion from $\mathcal{A}_0$ to $\mathcal{A}$, $f^* N$ coincides with the functor $i^* N$ defined in 3.4.

3.6. *Theorem: Let $\mathcal{A}, \mathcal{B}$ be small categories, $f: \mathcal{B} \to \mathcal{A}$ a functor, and N: $\mathcal{B} \to \mathcal{M}$ a functor from $\mathcal{B}$ to an abelian category $\mathcal{M}$ with exact infinite direct sums. There is a spectral sequence*

$$E^2_{p,q} = \varinjlim\nolimits_p^{\mathcal{A}}((L_q f^*) N) \Rightarrow \varinjlim\nolimits_{p+q}^{\mathcal{B}} N,$$

where $\varinjlim_p^{\mathcal{A}} L \simeq H_p(\mathcal{A}, L)$, $\varinjlim_n^{\mathcal{B}} N \simeq H_n(\mathcal{B}, N)$ and $L_q f^*$ are respectively the left satellites of $\varinjlim L$, $\varinjlim N$ and f^*.

Consider, in fact, the commutative triangle

$$\begin{array}{ccc} \mathcal{B}\mathcal{M} & \xleftarrow{f_*} & \mathcal{A}\mathcal{M} \\ & {}_{\Gamma_{\mathcal{B}}}\nwarrow \quad \nearrow_{\Gamma_{\mathcal{A}}} & \\ & \mathcal{M} & \end{array}$$

where $\Gamma_{\mathcal{A}}$ and $\Gamma_{\mathcal{B}}$ associate with $m \in \mathfrak{Ob}\,\mathcal{M}$ the constant functors from $\mathcal{A}$ and $\mathcal{B}$ to $\mathcal{M}$ with value m. The functor $\varinjlim^{\mathcal{B}}$, left adjoint to $\Gamma_{\mathcal{B}}$, is isomorphic to the composition $\varinjlim^{\mathcal{A}} \circ f^*$. In order to get the usual spectral sequence of a composite functor (Cartan-Eilenberg, XVI, 3 or Grothendieck), we have only to verify that:

(i) the left satellites of f^* exist;

(ii) for each $N \in \mathfrak{Ob}\,\mathcal{B}\mathcal{M}$, there is an epimorphism $p: N' \to N$ such that $\varinjlim_n^{\mathcal{B}} N' = 0$, $(L_n f^*)\, N' = 0$ and $\varinjlim_n^{\mathcal{A}} (f^* N') = 0$ if $n > 0$. This is done in 3.7 below.

3.7. It is a well known fact (see Röhrl) that, in order to prove the existence of the satellites $L_n f^*$, it is sufficient to construct a big enough family of f^*-acyclic objects of $\mathcal{B}\mathcal{M}$, i.e. a family $\mathcal{F}$ satisfying a) and b):

a) For each exact sequence $0 \to N' \to N \to F \to 0$ of $\mathcal{B}\mathcal{M}$, such that $F \in \mathcal{F}$, the sequence $0 \to f^* N' \to f^* N \to f^* F$ is exact.

b) For each exact sequence $0 \to N' \to N \to N'' \to 0$ of $\mathcal{B}\mathcal{M}$, there is an exact commutative diagram (Fig. 114)

$$\begin{array}{ccccccccc} 0 & \to & F' & \to & F & \to & F'' & \to & 0 \\ & & \downarrow & & \downarrow & & \downarrow & & \\ 0 & \to & N' & \to & N & \to & N'' & \to & 0 \\ & & \downarrow & & \downarrow & & \downarrow & & \\ & & 0 & & 0 & & 0 & & \end{array}$$

Fig. 114

with $F', F, F'' \in \mathcal{F}$.

If $j: \mathcal{B}_0 \to \mathcal{B}$ is the inclusion functor (see 3.5), we may choose for $\mathcal{F}$ the family $(j^* P)_{P \in \mathcal{B}_0 \mathcal{M}}$. Indeed, it is easy to verify that, if $a \in \mathfrak{Ob}\,\mathcal{A}$ and if $k: (f/a)_0 \to (f/a)$ is the inclusion, the square

$$\begin{array}{ccc} \mathcal{B}_0\mathcal{M} & \xrightarrow{\pi_*} & (f/a)_0\,\mathcal{M} \\ {\scriptstyle j^*}\downarrow & & \downarrow{\scriptstyle k^*} \\ \mathcal{B}\mathcal{M} & \xrightarrow[pr_{1*}]{} & (f/a)\,\mathcal{M} \end{array}$$

is commutative $(\pi: (f/a)_0 \to \mathcal{B}_0$ is induced by $pr_1: (b, \xi) \rightsquigarrow b)$. This proves a) by 3.4, if we notice that, for each $a \in \mathfrak{Ob}\,\mathcal{A}$, $(f^* N)\, a$ is equal to $\varinjlim\, (pr_{1*} N)$. The condition b) follows directly from 3.4 and the functoriality of the construction $N \rightsquigarrow j^* N$.

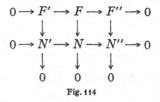

This proves condition (i) of 3.6. In order to prove (ii), it remains to prove that $\varinjlim_n^{\mathscr{A}} f^*(j^* P) = 0$ if $n > 0$ and $P \in \mathscr{B}_0\mathscr{M}$. This follows from 3.4 and from the commutativity of the diagram

$$\begin{array}{ccc} \mathscr{A}_0\mathscr{M} & \xleftarrow{f_0^*} & \mathscr{B}_0\mathscr{M} \\ {\scriptstyle i^*}\downarrow & & \downarrow{\scriptstyle j^*} \\ \mathscr{A}\mathscr{M} & \xleftarrow{f^*} & \mathscr{B}\mathscr{M} \end{array}$$

which itself follows from the equality $fj = if_0$.

3.7. Remark: The simplicial objects $C^*(\mathscr{A}, L)$ of 3.2 may be given a more sophisticated presentation with the help of the standard constructions of GODEMENT-HUBER: thus, if $i: \mathscr{A}_0 \to \mathscr{A}$ is the inclusion (3.5), it turns out that the standard construction of $\mathscr{A}\mathscr{M}$, which is associated with the trivial construction of $\mathscr{A}_0\mathscr{M}$ and the pair of adjoint functors i_* and i^*, gives rise to $\varinjlim$-acyclic resolutions in $\mathscr{A}\mathscr{M}$. The functor $\varinjlim^{\mathscr{A}}: \mathscr{A}\mathscr{M} \to \mathscr{M}$ maps these resolutions onto the differential objects $C_*(\mathscr{A}, L)$.

3.8. Remark: Consider the commutative square

$$\begin{array}{ccc} \mathscr{B}\mathscr{M} & \xrightarrow{f^*} & \mathscr{A}\mathscr{M} \\ {\scriptstyle pr_{1*}}\downarrow & & \downarrow{\scriptstyle ?a} \\ (f/a)\,\mathscr{M} & \xrightarrow[\varinjlim^{(f/a)}]{} & \mathscr{M} \end{array}$$

where a is an object of $\mathscr{A}$ and $?a$ is the evaluation functor $L \rightsquigarrow La$. In view of what has been proved in 3.6, we may calculate the left satellites of f^* $(?a) \circ f^*$ with the help of resolutions by objects of the form j^*P; as $?a$ is an exact functor, $\big((L_q f^*)\, N\big)\, a$ coincides with the value on N of the q-th satellite of $(?a) \circ f^* = \varinjlim^{(f/a)} \circ pr_{1*}$. As pr_{1*} is exact and transforms j^*P into a $\varinjlim$-acyclic object, this last value coincides with $\varinjlim_q^{(f/a)}(N\, pr_1)$.

This generalizes the Kan-construction, recalled in 3.5, since it shows that $L_q f^*$ may be calculated with the help of the q-th satellites $\varinjlim_q^{(f/a)}: (f/a)\,\mathscr{M} \to \mathscr{M}$ of $\varinjlim^{(f/a)}$.

4. The Spectral Sequence of a Fibration

4.1. We want first to show a connection between the homology groups of a simplicial set and the constructions of the preceding paragraph:

Let X be a simplicial set and consider the functor $p: \Delta/X \to \Delta$, which maps an object $\Delta[m] \xrightarrow{\tilde{x}} X$ of Δ/X onto $[n] \in \mathfrak{Ob}\,\Delta$ (II, 1.1). If p°: $(\Delta/X)^\circ \to \Delta^\circ$ is the functor induced by p, and if $[n] \in \mathfrak{Ob}\,\Delta^\circ$, the objects of $p^\circ/[n]$ (3.5) are the pairs $(\tilde{y}, \mu)$ formed by a singular simplex $\tilde{y}: \Delta[m] \to X$ $(y \in X_m)$ and a morphism μ: $[n] \to [m]$ of Δ. It follows that there is a unique morphism from $(\tilde{y}, \mu)$ to $(\tilde{y} \circ (\Delta\mu), \mathrm{Id}[n])$; the map, which associates with $x \in X_n$ the connected component of $p^\circ/[n]$ containing $(\tilde{x}, \mathrm{Id}[n])$, is a bijection of X_n onto the set of connected components of $p^\circ/[n]$ (a connected component is a maximal connected full subcategory); moreover $(\tilde{x}, \mathrm{Id}[n])$ is a final object of the connected component containing it. It then follows that if $F: p^\circ/[n] \to \mathcal{M}$ is a functor, $\varinjlim F$ may be identified with the direct sum $\coprod_{x \in X_n} F(\tilde{x}, \mathrm{Id}[n])$.

The preceding considerations show that, for any functor $L: (\Delta/X)^\circ \to \mathcal{M}$, the simplicial object $p^{\circ *} L: \Delta^\circ \to \mathcal{M}$, which will also be written $C_*(X, L)$, is given by

$$C_n(X, L) = \coprod_{x \in X_n} L\tilde{x}.$$

Moreover, the restriction of the face operator $d_i^n: C_n(X, L) \to C_{n-1}(X, L)$ to the direct summand $L\tilde{x}$, is the composition

$$L\tilde{x} \xrightarrow{L\alpha} L(\widetilde{d_i^n x}) \xrightarrow{in_{d_i x}} C_{n-1}(X, L),$$

where α stands for the following morphism of Δ/X

$$\begin{array}{ccc} \Delta[n-1] & \xrightarrow{\Delta(\partial_n^i)} & \Delta[n] \\ & {\scriptstyle \widetilde{d_i^n x}} \searrow \quad \swarrow {\scriptstyle \tilde{x}} & \\ & X & \end{array}$$

We again set $\delta_n = \sum_{i=0}^{i=n} (-1)^i d_i^n$, thus obtaining a differential object of $\mathcal{M}$, whose homology objects will be written $H_n(X, L)$. If $\mathcal{M} = \mathcal{Ab}$ and if L is the constant functor of value $\mathbb{Z}$, we have the equality $H_n(X, L) = H_n X$ (1.1). More generally, if $\mathcal{M}$ is arbitrary and if L is a constant functor of value $M \in \mathfrak{Ob}\,\mathcal{M}$, we will write simply $H_n(X, M)$ instead of $H_n(X, L)$.

4.2. *Proposition: Let X be a simplicial set and L: $(\Delta/X)^\circ \to \mathcal{M}$ a functor, where $\mathcal{M}$ is an abelian category with exact infinite direct sums. For any n, there is a natural isomorphism from $H_n((\Delta/X)^\circ, L)$ onto* $H_n(X, L)$ (see 3.2 and 4.1 for notations).

As the functor $p^{\circ *}$: $(\Delta/X)^\circ \mathcal{M} \to \Delta^\circ \mathcal{M}$ is exact by 4.1, it follows from 3.6 that $H_n((\Delta/X)^\circ, L)$ can be identified with $H_n(\Delta^\circ, p^{\circ *} L)$, so that we have to prove the following: if C is a simplicial object of M, and if we set as usual $\delta_n = \sum_{i=0}^{i=n} (-1)^i d_i^n$, the n-th homology object $H_n C$ of (C, δ_*) may be identified with $\varinjlim_n C$.

This, however, is a well known fact (by DOLD-PUPPE, it is equivalent to an even simpler assertion on differential objects of $\mathcal{M}$): first of all, $H_0 C$ coincides with $\varinjlim C$, since the diagram $\partial_1^0, \partial_1^1$: $[0] \rightrightarrows [1]$ of Δ is coinitial in Δ (i.e. for any functor F with domain Δ, $\varprojlim F$ coincides with the inverse limit of the diagram $F\partial_1^0, F\partial_1^1$: $F[0] \rightrightarrows F[1]$). Moreover, the functors $H_n ?$ give rise to an exact connected sequence of functors. For this reason, we have only to prove the existence, for any $L \in \mathfrak{Ob}(\Delta^\circ \mathcal{M})$, of an epimorphism q: $L' \to L$ such that $\varinjlim_n L' = H_n L' = 0$ if $n > 0$. The notations being those of 3.4, set $\mathcal{A} = \Delta^\circ$. If we choose for L' a functor $i^* N$, where N is just a family $(N_n)_{n \in \mathbb{N}}$ of objects of M, we have only to prove that $H_n L' = 0$ (3.4). Replacing all the N_n by 0 with the exception of one, we may suppose that $N_n = M$ for some n, and $N_m = 0$ if $m \neq n$. In that case, $(i^* N)_m$ is the direct sum of $\Delta([m], [n])$ copies of M, and $H_m(i^* N)$ is equal to $H_m(\Delta[n], M)$, the notations being those of 4.1. The simplicial set $\Delta[0]$ being a retract by deformation of $\Delta[n]$ (IV, 1.2), the proposition follows from Lemma 4.3 below.

4.3. *Lemma: Let f, g: $X \rightrightarrows Y$ be homotopic maps between simplicial sets. For any object M of $\mathcal{M}$, the map $H_n(f, M)$, induced by f, coincides with $H_n(g, M)$.*

The proof is similar to that given in 1.4. See also 4.8 below.

4.4. Let f: $X \to Y$ be a morphism of $\Delta^\circ \mathcal{E}$ and Δ/f: $\Delta/X \to \Delta/Y$ the functor which maps $\Delta[n] \xrightarrow{\tilde{x}} X$ onto $\Delta[n] \xrightarrow{f\tilde{x}} Y$. We will give in 4.5 below a simple description of the functor $(\Delta/f)^*$: $(\Delta/X)\mathcal{M} \to (\Delta/Y)\mathcal{M}$ (3.5). On the other hand, we have given in 4.2 a simple description of the homology groups $H_n((\Delta/X)^\circ, L)$, where $L \in \mathfrak{Ob}((\Delta/X)^\circ \mathcal{M})$.

In order to apply § 3 by combining these two descriptions, we consider an arbitrary small category $\mathcal{A}$, and a functor L: $\mathcal{A} \to \mathcal{M}$, and suppose $L\alpha$ invertible for any $\alpha \in \mathfrak{Ar}\,\mathcal{A}$. Define then L^{-1}: $\mathcal{A}^\circ \to \mathcal{M}$ by $L^{-1} a = L a$ if $a \in \mathfrak{Ob}\,\mathcal{A}$ and $L^{-1}\alpha = (L\alpha)^{-1}$ if $\alpha \in \mathfrak{Ar}\,\mathcal{A}$:

Proposition: (The notations and assumptions are those 3.3.) If L is invertible for any $\alpha \in \mathfrak{Ar}\,\mathcal{A}$, there are natural isomorphisms between the objects $H_n(\mathcal{A}, L)$ and $H_n(\mathcal{A}^\circ, L^{-1})$.

Indeed, with the notations of 3.2, $C_n(\mathcal{A}^\circ, L^{-1})$ is equal to $\coprod_\alpha L a_n$, where α runs through the n-sequences

$$a_n \xleftarrow{\alpha_n} a_{n-1} \cdots a_1 \xleftarrow{\alpha_1} a_0$$

of $\mathcal{A}$. It is easy to verify that the morphisms

$$(-1)^n \coprod_\alpha L(\alpha_n \ldots \alpha_1)\colon\ L a_0 \to L a_n$$

make up an isomorphism of differential objects from $C_*(\mathcal{A}, L)$ onto $C_*(\mathcal{A}^\circ, L^{-1})$.

4.5. We return now to a morphism $f\colon X \to Y$ of $\Delta^\circ\mathcal{E}$; we consider a singular simplex $y\colon \Delta[n] \to Y$ of Y and the pull-back $F_y = \Delta[n] \underset{Y}{\times} X$ of the diagram

$$\begin{array}{ccc} & & X \\ & & \downarrow{\scriptstyle f} \\ \Delta[n] & \xrightarrow{y} & Y \end{array}$$

Thus, y is an object of Δ/Y and $(\Delta/f)/y$ (4.4) is nothing but Δ/F_y.

Define a covariant (resp. contravariant) *local system on* X as a functor $L\colon \Delta/X \to \mathcal{M}$ (resp. $L\colon (\Delta/X)^\circ \to \mathcal{M}$), such that $L\alpha$ is invertible for any morphism α. By 4.4, 4.2 and the preceding remark, if L is a contravariant local system on X, the value on y of the n-th satellite $(L_n(\Delta/f)^*)L^{-1}$ is equal to $H_n(F_y, L|F_y)$, where $L|F_y$ is the composition

$$(\Delta/F_y)^\circ \xrightarrow{(\Delta/pr_2)^\circ} (\Delta/X)^\circ \xrightarrow{L} \mathcal{M}.$$

Suppose moreover *that f is a fibration*: in that case, we will see below that $(L_q(\Delta/f)^*)L^{-1}\colon y \rightsquigarrow H_q(F_y, L|F_y)$ is a covariant local system on Y. The associated contravariant local system will be denoted by $\mathcal{H}_q(f, L)$ and called *the local system of fibre homology*. By 3.6, we have proved:

Theorem: Let $f\colon X \to Y$ be a fibration of $\Delta^\circ\mathcal{E}$ and L a contravariant local system on X, with values in an abelian category with exact infinite direct sums. There is a spectral sequence

$$E^2_{p,q} = H_p(Y, \mathcal{H}_q(f, L)) \Rightarrow H_{p+q}(X, L),$$

the notations being those of 4.1.

4.6. We still have to prove the following: *if L is a contravariant local system on X, the functors $y \rightsquigarrow H_q(F_y, L|F_y)$ are covariant local systems on Y*. In fact, consider a morphism

$$\begin{array}{ccc} \Delta[n'] & \xrightarrow{t} & \Delta[n] \\ & {\scriptstyle y'}\searrow \quad \swarrow{\scriptstyle y} & \\ & Y & \end{array}$$

of Δ/Y and the induced morphism $H_q(F_t, L)\colon H_q(F_{y'}, L|F_{y'}) \to H_q(F_y, L|F_y)$. In order to prove that $H_q(F_t, L)$ is invertible, it is sufficient to look at the special cases $t = \Delta(\partial^i_n)$ and $t = \Delta(\sigma^i_n)$. In the first case, t has a retraction s such that ts is homotopic to $\mathrm{Id}\,\Delta[n]$; apply then VI, 5.4.2 with $E = F_y$, $B = \Delta[n]$, $A = \Delta[n-1]$; by 4.9 below $H_q(F_t, L)$ is invertible. In the second case, t has a section s such that st is homotopic to $\mathrm{Id}\,\Delta[n+1]$; apply then VI, 5.4.2 with $E = F_{y'}$, $B = \Delta[n+1]$, $A = \Delta[n]$.

4.7. In order to prove 4.9 below, let L be a contravariant local system on a simplicial set Y. We associate with L a local system ϱL on ΠY (appendix 1, 1.2), i.e. a functor $\varrho L\colon (\Pi Y)^\circ \to \mathcal{M}$ where ΠY is the

Poincaré groupoid of Y. If $x \in Y_0$, we set $(\varrho L)x = L\tilde{x}$, where $\tilde{x}$ is the singular simplex associated with x; it remains to associate with each $s \in Y_1$ an invertible morphism $(\varrho L)s\colon (\varrho L)(d_0 s) \to (\varrho L)(d_1 s)$, which is compatible with the relations given in II, 7.1. This is done by the formula $(\varrho L)s = (L\beta)(L\alpha)^{-1}$, where α and β are the following morphisms of Δ/Y

$$\begin{array}{ccc} \Delta[0] & \xrightarrow{\Delta\partial_1^0} & \Delta[1] \\ & {\scriptstyle \widetilde{d_0 s}}\searrow \quad \swarrow{\scriptstyle \tilde{s}} & \\ & Y & \end{array} \qquad \begin{array}{ccc} \Delta[1] & \xrightarrow{\Delta\partial_1^1} & \Delta[0] \\ & {\scriptstyle \tilde{s}}\searrow \quad \swarrow{\scriptstyle \widetilde{d_1 s}} & \\ & Y & \end{array}$$

Conversely, let P be a local system on ΠY. We define then a contravariant local system τP on X as follows: let $\eta_n\colon [0] \to [n]$ map 0 onto 0; for any $x \in X_n$, we define $(\tau P)\tilde{x} = P(X(\eta_n)x)$; for any morphism

$$\begin{array}{ccc} \Delta[m] & \xrightarrow{\Delta\varepsilon} & \Delta[n] \\ & {\scriptstyle \tilde{y}}\searrow \quad \swarrow & \\ & Y & \end{array}$$

of Δ/Y, let $\vartheta\colon [1] \to [n]$ be such that $\eta_n = \vartheta\partial_1^1$ and $\varepsilon\eta_n = \vartheta\partial_1^0$; we then define $(\tau P)(\Delta\varepsilon)$ to be $P\vartheta$.

It is easy to verify that the functors ϱ and τ are quasi-inverse to each other. Thus, *$\varrho\colon L \rightsquigarrow \varrho L$ is an equivalence of the category of contravariant local systems on Y onto the category of local systems on ΠY.*

This shows in particular that, *if Y is connected and simply connected, each contravariant local system on Y is isomorphic to a constant one.*

4.8. *Lemma: Let $f, g\colon X \rightrightarrows Y$ be homotopic morphisms of $\Delta^\circ\mathscr{E}$ and L a contravariant local system on Y, with values in $\mathscr{M}$. There is an isomorphism $i\colon f^{-1}L \cong g^{-1}L$ such that the triangle*

$$\begin{array}{ccc} H_q(X, f^{-1}L) & \xrightarrow{H_q(X,i)} & H_q(X, g^{-1}L) \\ & {\scriptstyle H_q(f,L)}\searrow \quad \swarrow{\scriptstyle H_q(g,L)} & \\ & H_q(Y, L) & \end{array}$$

is commutative for all q.

In this lemma, $f^{-1}L$ stands for the composition

$$(\Delta/X)^\circ \xrightarrow{(\Delta/f)^\circ} (\Delta/Y)^\circ \xrightarrow{L} \mathscr{M};$$

we write $H_q(f, L)$ for the morphism induced by f. Clearly, we can reduce the proof to the case where f and g are connected by a homotopy h: $f = h\varepsilon_0$, $g = h\varepsilon_1$. In that case, if $pr_2\colon \Delta[1] \times X \to X$ is the canonical projection, it is clear that the local systems on $\Pi(\Delta[1] \times X)$, associated with $h^{-1}L$ and $pr_2^{-1}\varepsilon_0^{-1}h^{-1}L$ (4.7), are isomorphic. Hence, there is an isomorphism of $pr_2^{-1}\varepsilon_0^{-1}h^{-1}L$ onto $h^{-1}L$, and we are reduced to the case where Y is equal to $\Delta[1]\times X$, $f=\varepsilon_0$, $g=\varepsilon_1$ and $L=pr_2^{-1}N$, N being a contravariant local system on X.

In this last case, $\varepsilon_0^{-1}L$ and $\varepsilon_1^{-1}L$ are both equal to N; we choose $i=\operatorname{Id} N$ and construct morphisms s_n: $C_n(X, N)\to C_{n+1}(\Delta[1]\times X, L)$ such that $\delta_{n+1}s_n+s_{n-1}\delta_n=C_n(\varepsilon_1, L)-C_n(\varepsilon_0, L)$ (4.1). In order to define s_n, notice that the $(n+1)$-simplices of $\Delta[1]\times X$ are the pairs (τ_i, y), where $y\in X_{n+1}$ and where τ_i is defined as in 1.4; notice also that $L(\tau_i, y)=Ny$. We define the restriction of s_n to the direct summand Nx of $C_n(X, N)$ as the alterning sum $\sum_{i=0}^{i=n}(-1)^i\chi_i$, where χ_i is the composition

$$N\tilde{x}\xrightarrow{N(\sigma_n^i)}N({}_Xs_i^n x)=L(\tau_i, {}_Xs_i^n x)\xrightarrow{\;in\;}C_{n+1}(\Delta[1]\times X, L),$$

in being the canonical monomorphism of index $(\tau_i, {}_Xs_i^n x)$. This generalizes the construction given in 1.4.

4.9. *Lemma: Let* $X\underset{v}{\overset{u}{\rightleftarrows}}Y$ *be morphisms of* $\Delta^\circ\mathscr{E}$ *such that* vu *and* uv *are homotopic to the identities of* X *and* Y *respectively. If* L *is a contravariant local system on* Y*, the morphisms*

$$H_q(u, L)\colon H_q(X, u^{-1}L)\to H_q(Y, L)$$

are isomorphisms for each q.

In fact, if uv is homotopic to $\operatorname{Id} Y$, we have by 4.8 a commutative diagram

$$\begin{array}{ccc} H_q(X, u^{-1}L) & \xrightarrow{H_q(u,L)} & H_q(Y, L) \\ & \nwarrow{\scriptstyle H_q(v,u^{-1}L)} & \wr\big\downarrow{\scriptstyle H_q(Y,i)} \\ & & H_q(Y, v^{-1}u^{-1}L) \end{array}$$

In the same way, if vu is homotopic to $\operatorname{Id} X$, there is a commutative diagram

$$\begin{array}{ccc} H_q(X, u^{-1}L) & & \\ {\scriptstyle H_q(X,j)}\big\uparrow\wr & \nwarrow{\scriptstyle H_q(v,u^{-1}L)} & \\ H_q(X, u^{-1}v^{-1}u^{-1}L) & \xrightarrow[H_q(u,v^{-1}u^{-1}L)]{} & H_q(Y, v^{-1}u^{-1}L) \end{array}$$

Therefore, $H_q(v, u^{-1}L)$ is an epimorphism and a monomorphism, hence an isomorphism inverse to $H_q(u, L)$.

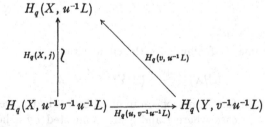

Bibliography

André, M.: Limites et fibrés. Compt. rend. **260**, 756—759 (1965).

— Derived functors in non-abelian categories (mimeographed notes).

Artin, M.: Grothendieck topology (mimeographed notes). Harvard 1962.

—, and B. Mazur: On the van Kampen theorem. Topology **5**, 179—189 (1966).

Barratt, M. G.: Track groups. Proc. London Math. Soc. (3) **5**, 71—106 (1955).

Barratt, M. G., V. K. A. M. Guggenheim, and J. C. Moore: On semi-simplicial fibre bundles. Am. J. Math. **81**, 639—657 (1959).

Cartan, H.: Séminaire E. N. S. 1956/57, exposé 1.

—, and S. Eilenberg: Homological algebra. Princeton University Press 1956.

Dold, A.: Die geometrische Realisierung eines schiefen kartesischen Produktes. Arch. der Math. **9** (1958).

—, and D. Puppe: Homologie nicht additiver Funktoren. Ann. Inst. Fourier Grenoble **11**, 201—312 (1961).

Eckmann, B., and P. J. Hilton: [*1*] Groupes d'homotopie et dualité. Compt. rend. **246**, 2444, 2555, 2991 (1958).

— [*2*] Transgression homotopique et cohomologique. Compt. rend. **247**, 620 (1958).

Eilenberg, S., and J. A. Zilber: [*1*] Semi-simplicial complexes and singular homology. Ann. of Math. **51**, 499—513 (1950).

— [*2*] On products of complexes. Am. J. Math. **75**, 200—204 (1953).

Epstein, D. B. A.: Semisimplicial objects and the Eilenberg-Zilber theorem. Invent. math. fasc. 3, **1** (1966).

Gabriel, P.: Catégories abeliennes. Bull. Soc. Math. France 1962.

Godement, R.: Théorie des faisceaux Hermann Paris 1958.

Grothendieck, A.: Sur quelques points d'algèbre homologique. Tohuku Math. J., Ser. II, **9**, 120—221 (1957).

Guggenheim, V. K. A. M.: On supercomplexes. Trans. Am. Math. Soc. **85**, 35—51 (1957).

Hilton, P. J.: Homotopy theory and duality. Cornell University 1959 (mimeographed lecture notes); Gordon and Breach 1965.

—, and S. Wylie: Homology theory, an introduction to algebraic topology. Cambridge University Press 1960.

Huber, P. J.: Homotopy theory in general categories. Math. Ann. **144**, 361—385 (1961).

Kan, D. M.: [*1*] Abstract homotopy I. Proc. Natl. Acad. Sci. U.S. **41**, 1092 (1955).

— [*2*] Abstract homotopy II, III, IV. Proc. Natl. Acad. Sci. U.S. **42**, 255, 419, 542 (1956).

— [*3*] A combinatorial definition of homotopy groups. Ann. of Math. **67**, 282—312 (1958).

— [*4*] Adjoint functors. Trans. Am. Math. Soc. **87**, 294—329 (1958).

— [*5*] On homotopy theory and C. S. S. groups. Ann. of Math. **68**, 38—53 (1958).

— [*6*] The Hurewicz theorem. Proc. Int. Symp. Algebraic Topology and its applications Mexico 1956.

— [*7*] On the homotopy relation for C. S. S. maps. Bol. Soc. Math. Mexicana **1957**, 75—81.

— [*8*] On C. S. S. complexes. Am. J. Math. **79**, 449—476 (1957).

KAN, D. M.: [9] On C. S. S. categories. Bol. Soc. Math. Mexicana 1957, 82—94.
— [10] Minimal free C. S. S. groups. Illinois J. Math. **2**, 537—547 (1958).
— [11] A relation between CW complexes and free C. S. S. groups. Am. J. Math. **81**, 512—528 (1959).
LAMOTKE, K.: Beiträge zur Homotopietheorie simplizialer Mengen. Bonn. Math. Schr. 17 (1963).
MACLANE, S.: [1] Simplicial topology. Lecture notes by J. YAO Chicago 1959.
— [2] Homology. Die Grundlehren der mathematischen Wissenschaften, Bd. 114. Springer 1963.
MILNOR, J. W.: [1] The construction FK. Princeton University (mimeographed) 1955.
— [2] The geometric realization of a semi-simplicial complex. Ann. Math. **65**, 357—362 (1957).
— [3] On spaces having the homotopy type of a CW-complex. Trans. Am. Math. Soc. **90**, 272—280 (1959).
MOORE, J. C.: [1] Semi simplicial complexes and Postnikov systems. Proc. Int. Symp. on algebraic topology and its applications. Mexico 1956.
— [2] Homotopie des complexes monoïdaux. Séminaire H. Cartan 1954/55.
— [3] Systèmes de Postnikov et complexes mono daux. Séminaire H. Cartan 1954/55.
— [4] Seminar on algebraic homotopy. Lecture notes. Princeton 1955.
— [5] C. S. S. complexes and Postnikov systems. Lecture notes Princeton 1957.
PUPPE, D.: Homotopiemengen und ihre induzierten Abbildungen I. Math. Z. **69**, 299—344 (1958).
RÖHRL, H.: Über Satelliten halbexakter Funktoren. Math. Z. **79**, fasc 3 (1962).
VERDIER, J.: Thesis to appear.
WHITEHEAD, J. H. C.: Combinatorial homotopy I. Bull. Am. Math. Soc. **55**, 213—245 (1949).
ZILBER, J. A.: Categories in homotopy theory. Dissertation (mimeographed) Harvard University 1963.
ZISMAN, M.: Quelques propriétés des fibrés au sens de Kan. Inst. Fourier Grenoble **10**, 345—457 (1960).

Index of Notations

Terminological Index